Impossible Horizon

Jacques Arnould

'All that is impossible remains

to be achieved.'

(Jules Verne)

Impossible Horizon

The Essence of Space Exploration

Jacques Arnould

Adelaide
2017

Published by:

An imprint of the ATF Press Publishing
Group owned by ATF (Australia) Ltd.
PO Box 504
Hindmarsh, SA 5007
ABN 90 116 359 963
www.atfpress.com
Making a lasting impact

Table of Contents

Preface   ix

Introduction: When the Horizon Fled   xi

1. Tsiolkovski's Dream . . . and Then?   1

2. We Still Went to the Moon!   9

3. A Space Oddity   13

4. 'Is the Sky Open to Us?'   17

5. Is Exploration Unique to Humans?   21

6. Need for Origin   25

7. Space at all Costs?   29

8. Heroism or Suicide?   35

9. Does the 'Envoy of Humankind' Concept Have a Future?   39

10. Avatars   45

11. A Virtuous Circle   53

12. Casting the Dye   63

13. The Earth does not Move   67

14. The Greater Earth   71

15. Plea for a New Horizon   75

References   79

Index of Names   81

Preface

Most children grow up by hearing their parents tell them fairy tales. For me, they were rather about exploration stories. Captivated, I discovered the stratospheric ascensions of my grandfather Auguste and the dives of my father and his mythical feat of descending into the Mariana Fosse with Don Walsh. The deepest point of the oceans!

The names I heard every day at home were those of Edmund Hillary, Charles Lindbergh, Alan Shepard, Wernher Von Braun, Jacques Mayol. They became the heroes of my childhood. But what really changed my life was to meet them all personally, to discover that it was not just beautiful stories, but a reality even more fabulous than all the fairy tales.

For me, exploration was the only way to live, and I was convinced that everyone shared that mindset: Coming out of certainties and habits to enter the unknown, doubts and uncertainty; using question marks to stimulate creativity and invent new solutions; to transform the impossible into possible! Is there another way to live his life? I thought not, until I realised that the explorer's state of mind was actually very scarce on this planet. The exploration of the unknown frightens all those who prefer to reassure themselves with dogmas, paradigms and common assumptions. What a disappointment . . .!

I am often asked how one becomes an explorer. In fact, you do not necessarily decide what you're going to explore, you just decide to get off the beaten track, take all the side roads, take every opportunity to do what others do not dare or 'Do not do'. Or rather what they think is impossible. Is that so for all explorers? I do not know, but that is how I lived my life, following the red thread of my childhood dreams. It was not north that indicated my inner compass, but the unknown.

It is not just about breaking records or performing spectacular actions. A record is only about beating the performance of the one who preceded us. The explorer is able to better: discover real novelties or perform firsts, that is to say accomplish something that nobody had yet succeeded or even believed possible. An explorer succeeds first, not just sets or makes records. And all the first explorers or record holders, who populated the narratives of my childhood, have been profoundly useful to humanity. They have opened up new routes, new modes of transport. They have changed the face of the world and, above all, profoundly changed our perception of the impossible. Those that have had no direct practical consequences, such as the conquest of the highest peaks, have had the capacity to give hope to humankind by showing to humanity what the human being is capable of accomplishing with courage and perseverance. Some have also been instrumental in protecting the environment. Like, for example, the diving of the Bathyscaphe in the Mariana Trench: by discovering a fish at 11,000 meters deep, Don Walsh and my father had been able to put an end to the government's plans to throw radioactive and toxic waste into the abyss that everyone thought was deserted.

Today, humanity faces major challenges. The challenges will open new horizons to science, but their goals will certainly be less to conquer unknown territories than to preserve the planet from the current threats to improve the quality of life. Let us dare to compare ourselves with the past: the state of humanity today resembles that of the planet before the great explorers began to explore it! There is still so much to do to reveal the hidden capital . . .

An impossible horizon, writes Jacques Arnould in this work, but a horizon without which our adventures, our explorations, would lose their savor and especially their meaning. We will then understand that even if the goal is never fully achieved, it is the quest that enriches us.

Bertrand Piccard

Bertrand Piccard is a Swiss psychiatrist and balloonist. In 1999, along with Brian Jones, he was the first to complete a non-stop balloon flight around the globe, in a balloon named Breitling Orbiter 3. He was the initiator, chairman, and co-pilot, with André Borschberg, of Solar Impulse, the first successful round-the-world solar powered flight.

Introduction
When the Horizon Fled

In those days, the Earth was flat, and round, like a plate. It was encircled by the Ocean river which seemed to go on forever; in its centre stretched the Mediterranean Sea that our ancestors, both affectionately and warily, called *Mare nostrum*, our Sea. The sky above it was a solid vault, like an over-turned salad bowl, studded with the stars. This was how the world appeared to the eyes and in the imaginations of our ancestors. It was a world on their scale, that is, on the level of their biped line of vision, on the scale of the trees they still climbed easily, the hills they ascended and the structures they built. Since it was flat, the Earth fit within human horizons. To explore it, all humans needed to do was put one foot in front of the other again and again without fear or tiredness. Curiosity would take care of the rest.

In the seventh century BC, as the Babylonians were building the highest tower yet, the one which would be known as the 'Tower of Babel', standing ninety metres tall, a handful of men appeared in Greece—philosophers, surveyors and astronomers. They were not satisfied with what they could see with their eyes. They wanted to know the how and why of the world and understood that they could only acquire this knowledge by combining their intelligence with another extraordinary ability that nature had given to humans: imagination, this capacity to distance oneself from immediate reality, to take the mind to another place far beyond. It was without hesitation that they conceived a spherical Earth, the antipodes, and the celestial bodies orbiting around the Earth and passing beneath it. The world spilled outside its former boundaries, outside human horizons, hiding behind its own roundness. Our ancestors therefore understood that in order to discover these unknown places, to go beyond the changing borders of their horizons, they would need to do more than sail the seas; they would also need to climb into the sky, to attain the domain of the birds, and maybe even the domain of the gods. Engineers and scholars joined adventurers in their exploration of the Earth.

For a long time, what we now know as 'space' was inaccessible and off-limits to humans, not because it was at a height which was unattainable without the least astronautical technology or principles, but because of the cosmic and dualistic representation of reality. Reality was seen as a nice, orderly whole, which the Greeks specifically named 'cosmos', within which humans were relegated to the centre—to a sort of 'cesspit' of imperfection, alteration, incompleteness and finally death. Around them were crystal spheres which held the planets and stars—immutable, eternal and perfect—a domain which was completely off-limits to humans, unless they had discarded their carnal envelope, either through a mystical experience or after death.

It took a revolution, what we now call the Copernican Revolution, to shatter the celestial spheres and make them no longer forbidden territory. Galileo was one of the first revolutionaries: through his astronomical observations, he showed that the Earth and the Sky were in fact made of the same fabric, the same material, and therefore belonged to the same world. He argued for the unification and equalization of the universe, its content and its laws. Johannes Kepler, having read Galileo's observations and findings in a work entitled *Starry Messenger* which Galileo had sent him in April 1610, decided to give the latter his full-fledged support, and penned his *Conversation with the Starry Messenger* in just eleven days. He wrote:

> There will certainly be no lack of human pioneers when we have mastered the art of flight. Who would have guessed that navigation across the vast ocean is less dangerous and quieter than in the narrow, threatening gulfs of the Adriatic, or the Baltic, or the British straits? Let us create vessels and sails appropriate for the heavenly ether, and there will be plenty of people unafraid of the empty wastes. In the meantime, we shall prepare, for the brave sky-travellers, maps of the celestial bodies—I shall do it for the Moon and you, Galileo, for Jupiter.[1]

---

1. Quoted in Arthur Koestler, *The Sleepwalkers: A History of Man's Changing Vision of the Universe* (London: Pelican, 1968), 378.

Intoxicated with the idea that mankind could one day escape its terrestrial prison, the 'little dungeon' described by Pascal, Kepler was convinced that now nothing was too high or too far for humanity to set its sights on reaching it. Rather than concerning himself with how to construct these vessels of the sky, Kepler left the task of inventing the art of flying to his engineers, and devoted himself to producing the maps which would be used by the first sky travellers. He saw the astronomer's work of map-making as essential, in order to distinguish and discover the worlds and islands, the reefs and sandbars that the *conquistadors* of space would encounter in their travels. Thus, the work of astrologers and soothsayers came to an end: it was no longer a question of reading the sky for fates and fortunes, or worse, the punishment awaiting intruders into the domain of some celestial power; it was a matter of assisting those who would one day attempt to write the destiny of humanity in the sky—the future explorers of space.

Centuries passed, and humans conquered air, and then space. Their feet touched the surface of the Moon and their wheels touched the surface of Mars. They acquired the vision of the gods; they peered into an abyss almost thirteen billion light-years deep. The Earth and, with it, the entire universe somehow became flat again with no folds, no curves, at least in appearance, to hide any dark corners. The horizon once again retreated out of reach taking with it perhaps the last dreams of exploration. Not so. Twenty-first century humans experienced the same strange feeling of suffocation described by Pierre Loti in his *Roman d'un Spahi* written in 1881 describing a young mountain dweller who, sent to Senegal by the French government as a soldier, discovered the desert: 'yet the endless level always ended by saddening him and weighing on his imagination, accustomed to the hills; he felt a craving to go on and on, farther and farther, as though to widen his horizon, to see what was beyond.'

Loti is right: the human imagination does not like horizons which are too flat, too clear; humanity needs to meet resistance, brakes, constraints to stop them in their tracks, to cross them and to lead humans to new *terrae incognitae*, to new unknown territories. Does our actual experience of reality, from the subatomic to the astronomical scales, possess sufficient reliefs to which our imagination can encounter or is it on the contrary too flat to stimulate it? Is it possible that the powerful spring of exploration could miss one day to humanity?

1
Tsiolkovski's Dream . . . and Then?

I do not intend to describe here how humans have perceived space since the dawn of time: there is evidence that all cultural groups in successive epochs have been intensely fascinated by the heavens, even though philosophical and religious notions that the sky was out of reach of mortals on Earth held sway. There are innumerable cosmic mythical narratives about the heavens, that don't always make a clear distinction between what we would now call the atmosphere and space. We need only think of the famous myth of Icarus and Daedalus, which tells the story of the first air crash.

Apart from Icarus, the first written trace of space as imagined by humans seems to be a text by Lucian of Samosata, dating from about 180 AD and ironically entitled *The True History*, in which Lucian warns his readers that 'I'm going to tell you about things I've never seen, adventures that didn't happen to me and that no one else has told me; I shall add events that never happened and couldn't ever happen; so I recommend my readers not to believe a single word.' A pretty astonishing disclaimer, is it not? Lucian then relates how a ship was lifted into the sky during a mighty tempest, travelled through the air for seven days and seven nights, arrived on a mysterious cosmic island, encountered strange creatures, and so on. In a different work, the *Icaromenippus*, Lucian tells the story of another voyage of discovery, by the eponymous Menippus, a man who has given himself wings to take him to the Moon and beyond.

Since Lucian, many authors have explored this literary genre. At the beginning of the fourteenth century, Dante Alighieri proposed a tour of the seven spheres of the cosmos, guided by Beatrice. In 1638, Francis Godwin published his vision of a wildly enchanting lunar nature and of a people more successful than our own, in *The Man in the Moon*. Around the same time, Cyrano de Bergerac published *Les Etats et Empires de la Lune* (The states and empires of the Moon, 1657) followed by *Les Etats et Empires du Soleil* (The states and empires of the Sun, 1662). In 1765, Marie-Anne de Roumier published the seven volumes of *Voyages de Milord Céton dans les sept planets* (Lord Seton's Voyage Among the Seven Planets): a true astronomical epic. In 1835, Edgar Poe sent *Hans Pfaal* to the Moon on board a balloon.

By the end of the nineteenth Century, the West had completed the systematic exploration of the Earth that it had been carrying on since the end of the fifteenth Century. Where could mankind now turn to satisfy its curiosity, its thirst for exploration? To space, of course; henceforth, space travel ceased to be purely imaginary and became a scientific possibility. Astronomers could use telescopes, first created in the seventeenth Century, as a 'vehicle' for searching beyond the atmosphere: with their help, and sometimes with a generous pinch of imagination, astronomers examined the Moon, the planets and the stars. They were able to map the surface of Mars as accurately as they could map the Earth—or so they claimed; Angelo Secchi, Giovanni Schiaparelli and Percival Lowell even traced extraordinary systems of canals. By the end of the nineteenth Century the leading French astronomer and prolific writer, Camille Flammarion, the author of *La Pluralité des mondes habités*, wrote in *Les Merveilles Célestes* that 'nothing singles out the Earth as being the only world in the Solar System fit for habitation, and that, astronomically speaking, the other planets are just as capable of supporting life'. A fairly clear invitation to his contemporaries to reach out and explore new worlds.

The twentieth Century, too, lacked neither authors nor imagination: from HG Wells (*The War of the Worlds* in 1898 and *The First Men in the Moon* in 1901) to Arthur C Clarke (*2001, Space Odyssey* in 1945), many writers mined the seam of imaginary space exploration. Wells, in fact, is sometimes said to have invented modern science fiction; it was he who brought such 'scientific marvels' as time machines, the device for transmuting matter and even hyperspace into common currency. For his part, Clarke, a member of the *British Interplanetary Society*, published an article in 1939 entitled '*We* Can Rocket to the *Moon*-Now!' before going on to work out the concept of geostationary satellites. The imaginary worlds that spring from the fertile minds of the authors of science fiction and 'space operas' seem as infinite as the Universe; and, not to be outdone, the cinema soon followed in the footsteps of literature; in 1902, Georges Méliès's masterpiece, *Le Voyage dans la Lune*, made use of all the latest technology.

Other cultures, of course, have their own versions of this process. I recall an official with responsibility for Japanese space policy explaining that his country's fascination with lunar missions could be attributed to their fondness for the old folktale about the Princess Kaguya. This was a mysterious girl apparently from the Moon who was found inside a bamboo stalk and was given the name *Naotake no Kaguya Hime*, 'the radiant princess of the bamboos'; she married a great lord, an earthling, but to her husband's dismay was carried back to the Moon by her father.

This possible influence of the imaginary aspects of a country's culture on its commitment and policy regarding space is clearer still in the case of the USA, to the extent that it is no exaggeration to refer to an *American space dream*. The American historian Howard McCurdy has shown in that, since its very beginnings, the American space programme was driven by a particularly romantic dream.[1] His theory is that more interesting because North American culture does not have the same fertile literary or artistic background as Europe; for most Americans, Ovid and Lucian of Samosata, Cyrano de Bergerac and Bernard de Fontenelle, too far back in time, are quite unknown or at the very least, have no real influence. Although they honour the memory of Lewis and Clark, who led the first American expedition across the continent to the Pacific at the beginning of the nineteenth Century, the Americans of the first half of the twentieth Century thought of the heavens as a phenomenon to be observed rather than a place one could travel to. They were familiar with the stories of Jules Verne and his imaginary technical world; but McCurdy suggests that they seemed mistrustful of his ideas. The United States could hardly be described as a nation of Luddites, but its people seem to have some difficulty in grasping the possibilities offered by technology. With the result that space travel in fiction is often shown as being much easier than it really is. Nor did it help that Robert Goddard, the father of American rocket science, was not a gifted communicator. In the end, however, after the Great Depression of the 1930s and the Second

---

1. See Howard E McCurdy *Space and the American Imagination* (Washington & London: Smithsonian Institution Press, 1997).

World War, things began to change. Certainly, the success of Soviet Russia's Sputnik satellites left the American government with little choice. It was a blow to their—claimed? supposed? genuine?—technological and military supremacy, and the United States needed to respond without delay. The start of the 1950s prepared public opinion for the launch of ambitious programmes and generated an interest in space travel. We should not underestimate the influence on public opinion of the writings of Willy Ley, the illustrations of Chesley Bonestell, the lectures of Wernher von Braun, and the films of Walt Disney. And we should not forget the groundswell of fascination with flying saucers and encounters with little green men, especially following the sighting reported by Kenneth Arnold, in June 1947: interplanetary voyages and extraterrestrial creatures were the two pillars of the American space dream. While the military rapidly gave up any interest in manned space flight, NASA made it one of its principal objectives, encouraged by President Kennedy's speech in 1962: from very early on, America's conception of space exploration shimmers with the silvery space suits of the seven astronauts of the Mercury programme, as they were presented to the press on 9 April 1959.

And the dream became reality, in a supreme effort which could be compared to a surrealist phenomena. In the preface to *Les Mamelles de Tirésias*, Guillaume Apollinaire wrote, 'When Man wanted to imitate walking, he created the wheel which does not resemble a leg. He did something surrealist without even knowing it.'[2] To fly, man had to implement a similar surrealism thousands of years later. To imitate the flight of birds, it wasn't enough to reveal the mystery of their wings and to build replicas as the flight of Daedalus very quickly confirmed! It was also important to observe windmills to invent the propeller. Propulsion in interstellar space would require even more effort, ingenuity and surrealism. To master it would once again be necessary to part from the obvious reality and accepted logic that is the entrapment of our Earth at the centre of ordered and eternal concentric crystal spheres, that of the said impossibility of moving in a vacuum. It took the works of Galileo and modern astronomers to challenge the deep-rooted conception of the cosmos, the works of Tsiolkovski and the first astronaut engineers to prove that the vacuum can be

---

2.   Quoted in Jean Brun, *Les conquêtes de l'homme et la séparation ontologique* (Paris: Presses Universitaires de France, 1961), 85.

travelled through in vessels built by man. There is nothing over the top therefore in the enthusiastic remarks of Hannah Arendt written in the foreword of her essay on *The Human Condition*. The German philosopher evokes the first Sputnik, the satellite imagined by human minds and built by human hands, which, in October 1957, circled Earth for several weeks 'according to the laws of gravitation that keep in motion the celestial bodies, the Sun, the Moon, the stars'. Sublime yet ephemeral company. An event that was 'second in importance to no other, not even the splitting of the atom', as it marked 'the first step toward escape from men's imprisonment to the earth',[3] writes Arendt who duly notes the dawning fulfilment of Tsiolkovski's dream. At the start of the twentieth century, when the foundations of modern astronautics were being laid, the Russian scholar wrote to one of his friends the words now known to all space aficionados: 'Earth is the cradle of Humanity; but mankind cannot stay in the cradle forever.'[4] At the end of the 1950s, all of the hopes usually associated with the discovery, conquest and exploitation of a new world were once again surfacing.

During the decade prophesied by Kennedy, space represented another upheaval similar to that which, for the Europe of the Renaissance, came with the discovery of a new world—with its riches, its dangers, and finally its inhabitants. The smile of Gagarin in April 1961, the incredible challenge set by the president of the United States and taken up by NASA, the technological and managerial odyssey of space programmes, the drama of Apollo 1, the consecutive 'premières' that resulted in the wishes of peace spoken by the Apollo 8 astronauts from lunar orbit in December 1968, man's first steps on the Moon, the bold diplomacy of the Apollo-Soyuz mission in July 1975: beyond the dream fed by spectacular images of the Earth seen from the Moon, by astronauts floating in nothingness then treading lightly on the Moon's surface, the first two decades of space history sparked and fuelled reasons to hope, in particular that of escaping the deep-rooted fears triggered by the Cold War. Were celestial bodies not declared the common heritage of humanity at the end of the 1960s? Did the signatory States of international space treaties not

---

3. Cf Annah Harendt, *Condition de l'homme moderne (The Human Condition)* (Paris: Calmann-Lévy, 1983), 33–34.
4. In a letter addressed to the engineer Boris Vorobiev on 12 August 1911.

agree to respect a peaceful use of space and not to militarise orbits? The age inaugurated by the feats of the first astronauts and cosmonauts gave us many reasons to hope, appearing to echo the words Tristan Bernard uttered as he was arrested by the Gestapo and quoted by François Jacob at the end of his work *Le jeu des possibles*: 'Till now we have lived with fear, now we can know hope.' Space in the 1960s indeed seemed full of hope.

What happened next? First came an almost immediate loss of interest in the missions that followed the triumphant Apollo 11 mission, except for Apollo 13, the mission that turned to disaster and narrowly avoided tragedy; then came the cancellation of the last two missions initially scheduled. On 14 December 1972, when the strange lunar spider module left the Valley of Taurus-Littrow, it was as if a small window closed in the night sky. It would be a long time before man left the confines of Earth again. We all know how the story unfolds. However, the next chapter was not disappointing. The space shuttle programme with its successes, its dramas and the obstinacy to perpetuate itself had all the ingredients of a new odyssey. The space station that today rotates over our heads is effectively international; the images gathered by the Hubble telescope fascinated the curious and even those less curious worldwide; the robotic missions on Mars, from Pathfinder to Curiosity, were followed by millions of internet users; the adventures of the Japanese probe Hayabusa, launched upon discovery of the Itokawa asteroid, inspired three Japanese films; the media success of the European probe Rosetta and its instrument Philae, to study the comet Churyomov Gerasimenko was unexpected. This wealth of exploration missions demonstrated the expertise, acumen, passion and tenacity of engineers. Space has not disappointed. It still and often captures the imaginations of those who learn about and discover new chapters in its history and its odyssey. However, I fear that the hope it once inspired has been lost. What can we really expect to be revolutionary about manned flights that seem confined to Earth's orbits for the next few decades and perhaps longer? What can we expect of a universe whose observable limits are restricted, even using our most advanced telescopes, lost in the mists of the unlimited and the infinite? Have we to consider the observation made by Pierre Auger, the first President of the French Space Agency (CNES): Many scientists, he wrote, would like to call for a sort of moratorium of science fiction, and tell science fiction writers: 'Hold it right there, stop giving away what we're going to do.' Every time scientists make a fresh breakthrough, the public are already used to the idea. They don't find new discoveries astonishing any more, and it's a pity? Is there really a future for space exploration or should we see the 1960s as a space oddity?

2
We Still Went to the Moon!

The feat of Neil Armstrong and Buzz Aldrin on 20 July 1969 will always be a key moment in the history of mankind and, because of this or despite this, can be used as a case study into exploration conditions. Not only did the two men walk on the surface of the Moon and take the largest step ever taken by humans, but their feat was watched by a television audience of over half a billion viewers, not including the many more listening in on the radio. Never before had a event of this kind and this scale unfolded in front of so many witnesses. This mainly technological innovation allowed man to illustrate and to embody the spirit of the treaty adopted in December 1966, 'on principles governing the activities of States in the exploration and use of outer space, including the Moon and other celestial bodies': article V states that astronauts are 'envoys of mankind'. In fact, after the return to Earth of the Apollo 11 crew, many earthlings proudly attested: 'WE went to the Moon!' Did Neil Armstrong foresee this reaction when, after having carefully descended the ladder of the lunar module and trodden on the fine grey dust, he spoke the unforgettable words: 'That's one small step for man, one giant leap for mankind.'?

When launching the eleventh mission of the Apollo programme, the NASA and United States' authorities no doubt considered that they had planned for every (or nearly every) eventuality, including the worst. A speech had even been pre-written by the American president Richard Nixon to be read in the event of a failure of the most dangerous phase of the entire mission, the firing of the engine of the lunar module transmission stage; an operation that had never been tested in lunar conditions. Similarly, scientists advocated a cautious quarantining of the crew following their sea landing in the Pacific Ocean. This safety measure was however quickly deemed to be a waste of time. However, even if all possible lunar expedition hazards had thus been planned for, all of the consequences associated with their success had not. Today, more than forty years after the last lunar mission, Apollo 17, took place, we should look beyond the immediate and spontaneous popularity with which Armstrong, Aldrin and Collins were met on their return from the Moon; we should challenge the repeated certainties and the evidence that was acquired too quickly. It is important to consider the doubts that are still raised today about the reality of lunar missions and their successes. If these suspicions and denigrations are in fact related to a deception similar to that which they claim to denounce, are they not at least telltale signs that the feat of the American astronauts would be affected, without anyone being accused of being the instigator or intentional cause, by the dual and apparently contradictory handicap of excessiveness and trivialisation?

Generally speaking, faced with the enormity of some feats, whether concurrent events or the complexity of systems giving rise to the events, with the enormity of intellectual genius, manual dexterity or individual courage demanded, it is not surprising that our minds begin to doubt and require explanations. By their nature, exploration missions abandon familiar shores and cross rational boundaries. Why are we surprised if former expeditions, mainly human (although it is important to remember the astronomic observations which, after the seventeenth century, themselves had consequences) like those of today (in which robots play an increasing role), raise doubts among those of us who cannot accompany explorers and who remain in a world that is constant and known to us? Too much novelty and unknown can make it difficult to accept for non-scholars and those left behind. It is true that explorers have always taken great pains to document their journeys, to prove the reality of their exploits and to prove the truth of their tales. Astronauts are no exception, feeding photographic and filmed images back to Earth from the Moon, sharing their adventures and bringing back carefully chosen stones and even dust trapped in the fibres of their spacesuits. Their exploits have also been relayed live to audiences via transistor radio waves, cameras and television screens. But nothing will do, not even the angry and desperate punch that Buzz Aldrin landed on one of the critics of this lunar exploit: NASA is still suspected today of having come up with a devilish plot, of having used the services of the Hollywood or Walt Disney studios . . . and of having forgotten that, on the real Moon, no breeze would have made the American flag wave! Let us leave these conspiracy theories to the experts and let them analyse these accusations, study historical deployment, understand motivations. They are bound to dig up food for thought in such an interesting case.

Among the reasons put forward to explain these doubts and suspicions and also these manipulations of public opinion, it is important to remember the following constant as evidence. The astronauts seem not only to have accomplished what the engineers and technicians had designed and taught them to do, but also what prophets and science fiction authors had been imagining, narrating and drawing for quite some time. We only have to look at the works of Jules Verne (*From the Earth to the Moon* and *Round the Moon*) and Hergé, the creator of Tintin (*Destination Moon* and *Explorers on the Moon*). When the reality of exploration blends with fiction and the imagination, it is the former that loses its charm and, due to its essential and inevitable excessiveness, suffers from irreversible trivialisation.

To explore this hypothesis further, it is important to read *Phases of Gravity* written by Dan Simmons and describing lunar missions; the book was published in 1989 and translated into French under the poetic title of *Les larmes d'Icare* (*Icarus' Tears*). Its hero, Richard Baedecker, had walked on the Moon. However, sixteen years later, he still felt unfulfilled and disillusioned:

> *It is just like the simulations,* he'd thought. Even during the final manoeuvre, he knew that he should have seen something else, and that he should have felt differently. He reacted on autopilot to the instructions transmitted by Houston, answered the technician's questions, entered the required data into the computer, repeated the figures to Dave, and during this time, the same depressing thought continued to ring in his mind: *Just like the simulations.*[1]

Despite the tragic and singular beauty of the loneliness he felt behind his helmet with the gold visor, his mission never seemed to him more than just another training mission at Houston and in the Meteor Crater, worst still, a repetition of the Apollo 11 mission. It even had the feeling of a poor replica of the lavish performances put together by talented artists several years before, such as Chelsey Bonestell whose paintings literally inspired the American space programme. In addition to Baedecker's encounters, Simmons thought it was important

---

1. Dan Simmons, *Les larmes d'Icare (Phases of Gravity)*. (Paris: Denoël, 1994), 16.

to introduce new friends, new loves, to restore the disparate pieces of his amazing adventure, to understand it and make it his own; in other words, to assimilate it. Before returning, brimming with experience and lucidity, to live the rest of his days on Earth . . . Would a feeling of deception, similar to that experienced by Baedecker, not go some way to explaining the doubts raised with regard to the Apollo missions? After discovering Stanley Kubrick's film *2001, A Space Odyssey*, which was released in cinemas in April 1968, six months before the first manned mission around the Moon, humans staying on Earth could have been disappointed when viewing the images transmitted to their television screens. And, even though most of us will never know what it is like to travel through space and time as the astronauts have done and recounted (oceans crossed in a matter of minutes, and a new sunrise every hour and a half), virtual reality and CGI can offer us very convincing substitutes that are almost as amazing and disorienting . . . unless we have become so familiar with the animated maps that illustrate weather forecasts, or Hollywood spy films, that we are blasé about it all and, like some modern airline passengers, prefer to pull a curtain over the porthole and plunge into a virtual world of entertainment and games.

Between excessiveness and triviality, between doubt and disillusion, perhaps it is important to raise the idea that the success of the Apollo missions came both too late and too soon. Too late, because human imaginations and minds had already seen and processed very detailed and exciting images of man reaching and walking on the Moon, so effective that the real event had little more impact or chance of wowing audiences. Too early because the same minds had not had enough time to understand the ins and the outs of the technological revolution of which the American (and not forgetting, the Soviet) exploits were simply the most effective; the tip of the iceberg remaining hidden and the rest of it still concealed in part today. Player but also informed observer of the astronautics scene, Arthur C Clarke did not hesitate to write, ten years after Apollo 11, that space travel 'is a technological mutation that should not really have arrived until the 21ˢᵗ century'.[2] If we consider Clarke's statement to be one of reason, perhaps mankind was not ready to go to the Moon in 1969.

---

2. Quoted in James A Dator, *Social Foundations of Human Space Exploration*. (New York: Springer-ISU, 2012), 27.

3

A Space Oddity

Chris Hadfield has been described as the most famous astronaut in the world; that is the most famous after the late Neil Armstrong. What exactly did do to gain such a status? During the six months prior to his return to Earth, on 14 May 2013, he became the first Canadian to captain the crew of the International Space Station. A few days before setting foot once again on terra firma, amongst the stunning steppes of Kazakhstan, he had supervised a delicate and urgent mission to repair the biggest space complex ever built. However, he owes his fame to the social network *Twitter*: 850,000 tweeters followed his time in space and his eighty videos went viral. The last video, in which he sings his version of David Bowie's *Space Oddity*, received more than fifteen million hits in just one month of being on line! At the same time, French secondary school students were unable to name Jean-Loup Chrétien as the first Frenchman to have gone into space in June 1982.

The aim is not to deny or devalue the exemplary qualities of these some 500 men and women who have left the terrestrial atmosphere and spent time in space. They continue today to explore both technological and anthropological boundaries, supported by the knowledge, power and curiosity of humans. It is down to them to report on their missions at these boundaries, as Colonel Hadfield did with such professionalism and brilliance. However, did the almost exaggerated success of Colonel Hadfield not, to the contrary, demonstrate the trivialisation or even total lack of interest with which most manned space missions are met today?

It is not even a question of regretting the Moon landing era. The risks undertaken by the astronauts still seem inordinate, especially given the undesirable socio-political context of the Cold War. The years went by, the iron curtain was lifted and the global geopolitical situation changed. The fierce battle between the Soviet Union and the United States subsided into a more rational balance creating competition and cooperation between the dozen space powers. It is impossible not to feel good about witnessing a Canadian astronaut singing and playing the guitar in the company of Russian, American, European and Japanese astronauts at an international space station! Even so, the return to Earth of the musical astronaut leaves a bitter aftertaste . . .

I didn't try to find out the reasons for Hadfield's artistic choice, which seemed somewhat odd. Released in 1969 several days before the lunar landing of the Apollo 11, *Space Oddity* tells the story of Major Tom and his communications with ground control. Following a textbook lift-off, the astronaut goes on a space walk. Shortly after, his craft experiences failure in space, like a common 'tin can'. Major Tom gradually loses contact with ground control and is forced to accept his fate: 'And there's nothing I can do' are the last words of the song . . . Performing these lyrics seems a strange way of bringing a long-term mission on board the International Space Station to a close. Why, I ask yet again, did Hadfield choose this song? Did he concede to the disappointment of having to leave his space and international tin can, to the nostalgia of space or, worse still, to despair? Of course it could just be that he simply chose a beautiful melody and a song which was an immediate hit at the time of its release. Regardless of the reason, the sign is there: one of the most famous astronauts (momentarily) in history returned to Earth after singing 'And there's nothing I can do'. In other words, it would seem that he brought nothing back from space other than a feeling of powerlessness when faced with a, fortunately totally imaginary, technical fault. It is not necessarily about satisfying human taste for adventure, but more so about confirming the thoughts of those who think that space venture has little to do now with the exploration movements which, in the past, marked the history of our species.

There are indeed many who criticise. They delight in recalling some of the things that were said during the International Astronautical Congress in Belgrade in September 1967. Projecting beyond the race to the Moon, even before Neil Armstrong and Buzz Aldrin had left the first human footprints on the surface of the Earth's natural satellite, the American and Soviet delegates had started to draw up plans to explore the Solar System. Once they had reached the Moon ('before the end of the decade'!), the next step would be a flyby of Venus in the early 1980s. The first astronauts or cosmonauts would land on Mars in 1983, would visit Venus in 1986 and, ten years later, Mercury. Before the end of the twentieth Century, according to these scientists and engineers, great exploratory spaceships would be going as far as Jupiter, and humans would obviously be exploring its moons. However, the visionaries at Belgrade did not think it reasonable to push their predictions any further than that.

In fact, wrote the French journalist Serge Brunier, the incredible exploration programme drawn up at Belgrade actually was accomplished—but by automated spacecraft, not people. 'Gradually,' he says, 'manned flights were abandoned by the stakeholders that had originally promoted space exploration, the military and scientists, who soon realised that humans in space were at best superfluous and at worst actually harmful'.[1] But, he went on to say that it did not stop NASA or ESA devoting a large part of their budgets to manned flights ('which, contrary to conventional wisdom, use the most rudimentary, trustworthy and conservative space technologies'). He then concludes: 'Astronauts are the surest and most efficient way of slowing down true space exploration.'[2]

These critics are an invitation to return to the anthropological roots of exploration and that of space in particular.

---

1. Serge Brunier, *Impasse de l'espace. À quoi servent les astronautes?* (Paris: Seuil, 2006), 18.
2. Brunier, *Impasse de l'espace,* 21.

4

'Is the Sky Open to Us?'

There are probably not many of us today who have read or even heard of the essay published by Walter Pons in 1960, entitled *Steht uns der Himmel offen?*—Is the sky open to us?.[1] This work is unique in that it is probably one of the first philosophical studies to focus on the astronautics venture, once this actually became reality. To the question which forms the title of this book, its author replies: 'We cannot really know the world if we do not firstly know ourselves.' This answer has undertones of Socrates as the famous Greek philosopher transformed the motto engraved on the front of the temple at Delphi and taught by priests: 'Know Thyself and leave the World to the Gods,' into a practically opposing formula: 'Know Thyself and you will know the Universe and the Gods.' In this teaching, Hegel saw a major change in human thinking. In effect, Socrates was proposing to make the inner conscience the human authority of truth and therefore of decision. It was no longer necessary, or even an option, to leave things to be sorted out by a superior and unattainable divine order, that could only be revealed by the oracle, by divination and mysticism. It was henceforth up to man to take his destiny in his own hands. Was this not precisely what twentieth century humans had attempted to do by committing to the risky conquest of space, by determining their own destiny instead of looking for it in the movement of the stars, like our predecessors? The question asked by Pons still remains to be properly answered: are humans, as individuals and as a species made to take on space, its immeasurable size, its hostility, without forgetting new

---

1. Cf Walter Pons, *Steht uns der Himmel offen? Entropie-Ektropie-Ethik. Ein Beitrag zur Philosophie des Weltraumzeitalters* (Wiesbaden: Krausskopf Verlag, 1960).

concerns about reality and new conceptions of the world it creates or even imposes?

As extreme and as emblematic, in terms of the question asked by Pons, seem experiences of extra-vehicular activities, in other words 'space walks'. Humans also come face to face with the most bizarre, the most grandiose and the most hostile aspects of space. All those who have travelled to space, and some on more than one occasion, will always remember their truly unique experience. French philosopher Maurice Blanchot, who passed away in 2003, clearly understood this when considering the significance of the first time man walked in space.

On 18 March 1965, protected by his spacesuit and attached to the Voskhod 2 by a solid lifeline, Alekseï Leonov became the first man to ever exit his vehicle and walk in space. 'I was moving towards the unknown and no-one in the world could tell me what I was about to find out there', he said:

> I had no instructions to follow. It was the first time. I knew it had to be done . . . I climbed out through the hatchway and carefully crawled out. I moved away slowly from the ship . . . What struck me most was the silence. It was a great silence, unlike any I have encountered on Earth, so vast and deep that I began to hear my own body [. . .] There were more stars in the sky than I had expected. The sky was deep black, yet at the same time bright with sunlight [. . .] The Earth was small, light blue, and so touchingly alone, our home that must be defended like a holy relic. The Earth was completely round. I believe I never knew what the word round meant until I saw the Earth from space.

As we know, Leonov's spacewalk was not without drama. Under the effect of internal pressure, his spacesuit started to inflate preventing him from bending his arms and legs. He initially refrained from telling his contacts on the ground but was unable to start up the camera he was carrying on his shoulder. After around twelve minutes, it was time for him to return to the Voskhod 2. However, the cosmonaut was unable to enter the airlock feet first, according to the planned procedure, and ended up entering head first. He was then forced to turn round in order to close the exit hatch which was impossible to do without opening a valve designed to lower the pressure in his spacesuit. Leonov managed to do this and, exhausted, rejoined his comrade and commander Pavel Beliaïev. However, that was not to be the only hiccough for the two men. An autopilot failure forced them to manually control their re-entry into the atmosphere and they touched down on Earth nearly four hundred kilometres from the scheduled landing point. They spent two nights in a snow-covered Siberian forest in fear of attack from wolves and bears. Leonov's first spacewalk thus ended with him skiing to meet the emergency helicopter!

This adventure inspired Blanchot to utter the following:

> . . . far away—in an abstract distance of pure science—removed from the common condition symbolized by the force of gravity, there was a man, no longer in the sky, but in space, in space which has no being or nature but is the pure and simple reality of a measurable (almost) void. Man, but a man with no horizon.[2]

The remark is neither too strong nor merely reduced to the symbolic. Man, who stepped across the parapet of his space balcony, man, who now walks in the void of space, is caught up in a dazzling circle around the Earth. Overcoming hills and mountains, pools and seas which, at the very bottom, obliterated his view and restricted his curiosity with limits and horizons, he only needed an hour and a half to circumnavigate the globe, to complete the tour and to examine the questions from all angles. Do we still have things to discover about the Earth, about ourselves, now that we as a species have reached this altitude, left the atmosphere, experienced zero gravity, danced with the stars? At an altitude of four hundred kilometres, the voices of the priests of Delphi and the voices of Socrates and Hegel can no longer be heard. Walter Pons was right to ask 'are humans ready for such an experience?'. Were they prepared to put such a great distance between the Earth and themselves, to brave the ultimate form of vertigo with no vertical or horizontal landmarks to guide them?

---

2. Maurice Blanchot, on the subject of the first extra-vehicular activity in space, quoted by Sophie Dupey in '*L'homme sans horizon*' (*The man with no horizon*), *narrative on the work of Richard Morice, painter*, http://sophie.dupey.over-blog. com/pages/l-homme-sans-horizon-texte-sur-le-travail-de-richard-morice-peintre-8326557.html.

Immediately following Apollo, the question was prudently deflected and forgotten. The Americans used the last Saturn rockets to install a space station around Earth; the Soviets did the same with their Salyut and Mir stations; the construction of the International Space Station required numerous extra-vehicular activities. The thriller, *Gravity*, directed by Alfonso Cuarón, spectacularly and dramatically depicted the unique experience of zero gravity around the Earth. However, the question asked by Pons remains open for discussion and Blanchot's analysis remains relevant. Confronted by the space vacuum, humans would now find themselves without horizon, having to set a new target, a new destination for their natural curiosity, a new propensity to explore. National and international space agencies and sometimes even private companies naturally talk of returning to the Moon, of reaching an asteroid, of landing on Mars ... but there are no concrete missions lined up as yet to show that these plans will ever see the light of day. It is as if the step to be taken has become too big for humanity, as if the horizon has escaped beyond the limits of our hopes, our dreams and our imaginations.

Should we have gone really to the Moon? Or should we have waited? The question is no longer relevant. The steps of Armstrong and Aldrin will be imprinted in the lunar dust for a long time. Humans returned to Earth. When the Apollo astronauts took a photo of our blue planet, a number of environmental movements emerged. Whilst Air France today promises to 'make the sky the best place on Earth', the French space agency, CNES, adopted the slogan: 'Space for the Earth.' The successors of the Sputnik do not take us far from our cradle and, on the contrary, help us to better take care of ourselves, to protect us from our own contamination and to preserve our planet for generations to come. 'Let us return to Earth; let us find intelligent horizons; let us think about our future and our children. Space exploration can wait': such slogans and arguments, now commonplace, are used in adverts and films. And so, was the conquest of space merely an interlude in the history of mankind, an episode to be forgotten to focus more clearly on humanity's uncertain future? I do not believe so and think it is important to re-establish the real roots of exploration.

5
Is Exploration Unique to Humans?

One of the explorers of the North Pole, Fridtjof Nansen explained 'The history of the human race is a permanent battle of light and darkness. It is not a question of the use to which the knowledge is put: man desires knowledge and when this desire dwindles, he is no longer a man.' His compatriot, Roald Amundsen, adopted a harsher approach: 'Little brains have only room for thoughts of bread and butter.' These two men viewed exploration and quest for knowledge as inseparable. In his work named *Leviathan*, Thomas Hobbes suggests that 'desire, to know why, and how, [is] curiosity'. Do the two explorers and the philosopher offer a satisfaying answer to the interrogation: is exploration unique to humans? If curiosity is one of the main driving forces behind the decision to explore and the actual act of exploring, and if it clearly goes hand in hand with the desire to know the why and the how, it would be reasonable for us to suggest that exploration is unique to humans . And we ready ourselves to deny non-human beings any attitude, any movement, that may hint at curiosity.

Although conceivable, this perspective results in delicate semantic convolutions: which term do we use to describe the daily behaviour of our family pets? What more do we need to qualify the behaviour of great tits studied by ornithologists who claim to be able to associate a particular form of encoding gene for dopamine receptor D4 with the ability of these birds to take an interest in an unusual object? Should we refrain from talking about their curiosity like ours? For the time being, I therefore propose not to follow the route mapped out by Hobbes, although I may well come back to that later, and to be satisfied with defining curiosity as interest, or simple availability, to a being or to a phenomenon.

Let's therefore start by looking at experience, which seems to be common to animals and to humans. Through the ordinary course of their lives, on the usual scene of their existence, a movement, a thing, a being, a person arrives who or which, straight away, seems to be unusual, strange, extraordinary. One of their senses raises the alarm. A dog's ears stand up, the eyes of a supposedly sleeping cat open, the great tit begins to stir, humans start. Our other senses quickly follow suit: what is happening? What is it? Who goes there? An unconscious movement, inherited or learned, this warning quickly turns into worry, even fear, causing us to defend ourselves, flee or attack. It can even provoke astonishment, surprise and, finally, curiosity. Is it surprising or shocking that these feelings, movements and reactions give rise to the possibility of providing grounds which scientists qualify as biological and philosophical aspects of natural existence? In other words, grounds common to both animals and humans. I am convinced that there is no reason to find this disturbing, quite the opposite. Having determined what we have in common, it is much easier to find out and understand our differences and our singularities. This brings us back to Hobbes.

Do we consider the cat which plays with its shadow or the shadow of its master to have a desire to know how this phenomenon works and also, to say the least, the experience of its inability to understand? Yes, of course. As unformulated as this question may be and remains to be, the cat's jumps and desperate paw swipes lead us to believe the rise of a type of confusion or annoyance that the friendly feline only displays when following the desperate moves of one of its prey. However, why this optical phenomenon occurs, if indeed it exists, probably does not fall within the cat's scope of thought. I believe that humans are currently the only species with the ability to ask such, almost metaphysical, questions.

I will not go into the many basic principles or reasons that can be associated with this human ability. I will just refer to the imagination, which needs to be firmly associated with curiosity. To sum up, it is because humans are able to move away from their immediate surroundings and to imagine themselves in another place and in another period of time, to step across the frontiers of space and time, that they ask and live with the question of why, the question of origins, whether those of a frightening shadow or their own. The child who bombards his parents with the eternal question: 'Mum, Dad, why . . .?' asks the question because he discovers they are able, due to their imagination, to remove themselves from an event or, conversely, immerse themselves in it. They also discover that they are able to imagine themselves in the place of someone else. This is the experience we as humans share when we ask ourselves about the why, that is, about the origin and purpose, sense and finality of our existence, beyond even the unique and forgotten experience of birth and about our inevitable death. Would we ask this if we had no imagination? We can only imagine (!) the answer. In any case, we see no signs of imagination in animals, even those of great intelligence.

The alliance of curiosity and imagination therefore seems to be unique to humans. In addition to our tendency to ask ourselves why, this alliance also drives us to explore. Has mankind never undertaken to explore worlds other than those he has previously imagined, of which he has already dreamed? Has he never explored territories other than those for which his maps have already shown the outlines, imagined the forms but leaving their content empty? The fascinating *terrae incognitae* of our ancestors, the breathtakingly far-reaching final frontiers of the modern world, which not only serve to fill in the space beyond the geographical horizon, to fuel the curiosity of the most adventurous humans, but also to offer snippets of the answer to the most nagging questions, those of our origins and our destiny, those of our identity and the possible existence of others. Exploration is therefore not a game to be played without asking why. It is one of the most serious human ventures of all time.

Curiosity, imagination and exploration form the knot that tightens around mankind's unique drama. However, we hold on to the sense of our most common and immediate nature. To put it another way, humans, and only humans, explore in the same way we breathe.

The birth of a child marks his acceptance into the circle of explorers: he is violently expulsed, torn from the original paradise, this Garden of Eden which has been his home for nine months. He is plunged into a world which until now could only be perceived through the horizon of his mother's rounded abdomen. In an instant, he discovers air, light, noises, smells, shocks, no longer protected inside his mother. And he begins to breathe, inhaling for the first time, exhaling for the first time. His pulmonary alveoli burst into life like the snap of sails blown by a storm. It is frightening and painful. A terrible pain, a harrowing cry, before the calm returns and he breathes unaided. This little one will not stop inhaling and exhaling until his death, breathing in and breathing out, concentrating and dilating the air necessary to his survival. And not just air: the world within and the world without, the self and the other, the known and the unknown continuously meet, clash with each other and jostle with each other throughout our lives at the gateways of our senses, at the counters of our knowledge, at the frontiers of our conscience. Yes, man explores like he breathes and breathes like he explores.

There is no longer any reason to be surprised by the etymology of the term 'to explore'. The prefix *ex* was added to the verb *plorare* (to cry, to weep), producing the original meaning of 'to cry out' or 'to scream'. But this seems to have little connection to what the term explore means to us today. What is even more perplexing is that adding the opposite prefix, *in,* produces a semiological result associated with the synergy that the difference, of the opposite between exterior and interior: *implorare* means to draw attention to oneself, to demand. It would without doubt be advisable to refer to a semantic derivation to explain the difference between the field of etymology and the usual usage of the term exploration; but is it useful? On the contrary, does this play with words and meanings not itself emphasise the inherently human character of the exploratory venture, which could never simply be reduced to the idea of doing something out of character and leaving one's territory; but still includes incorporation and ingestion? Exploration is and remains a dramatic venture inevitably filled with upset, cries and tears due either to happiness, sadness or suffering. Exploration remains a human venture, terrifyingly and magnificently human whose only outcome is death. *Usque ad mortem,* until death: such is the final end to human exploration.

6
Need for Origin

Has the time really passed when a wise man from the Bible could write, in the *Book of Proverbs*: 'There be three things which are too wonderful for me, yea, four which I know not: The way of an eagle in the air; the way of a serpent upon a rock; the way of a ship in the midst of the sea; and the way of a man with a maid.'? Now, we fix tiny transmitters to birds' wings to be able to track their majestic peregrinations; we have sophisticated devices that allow us to trace the slightest chemical substances left by reptiles in their environment; we observe the movements of boats and the sea via satellites; and we break down the most intimate biological processes using microscopes, scanners and ultrasounds. The world has therefore become transparent, exposed under man's watchful eye and instruments. To attain such a level of knowledge, sometimes even mastery, our predecessors, our contemporaries and we ourselves are not afraid to cross physical and psychological borders, to break spheres and taboos, to confront danger and take risks. To conquer what until now was too amazing for us, what taunted us. To reach and cross these terrestrial, maritime and aerial expanses evoked by the scholar. To acknowledge above all what we don't know: 'the way of a man with a maid', in other words the mystery of our birth and the mystery of our origin.

Circumnavigation of the globe undertaken by the young Charles Darwin, who began his journey on board the *Beagle* in December 1831 and finished six years later, was not a leisure trip offered by his father to congratulate him on completing his theology degree at Cambridge. Navigating on board his majesty's ships in the middle of the 19th century was very likely to turn into a true adventure and an amazing exploration expedition. Darwin recorded for all time entire areas of human knowledge about the living, their origins and their evolution, magnificently lifting the mysterious veil until then clouding their knowledge.

Several years after the famous English naturalist, anthropologists and prehistorians enquired about our first ancestors. 'The cradle of Humanity?' liked to joke one of these researchers, Father Henri Breuil ('l'Abbé Breuil'), 'Do not concern yourselves; this is a cradle on wheels!' These scholars took the route of the caves, entered the shadows where no-one had dared to venture for thousands of years, cave diving, discovering ancestral postures to reproduce fantastic paintings. The certainties to be shattered were no less than those criticised by their fellows Copernicus and Darwin. Religious minds took offence when they learned that they were not descended from an angel, itself fallen, but from a monkey, in the process of improvement and progress. Self-righteous minds refused to bestow the mastery of cave painting often as good as that in the Sistine Chapel upon their ancestors . . . Intellectual disputes in the shadows of academic cupolas; unintelligible confrontations with terrestrial meanderings; endless battles with relics of the past, using the hammer, pencil and tracing paper as weapons. All of these scholars, Darwin on board the *Beagle*, Breuil in the Lascaux caves, came up against unknown lands around the globe at the same time as the most sensitive areas of human knowledge: that of its origins.

What if one of the strongest, most certain, most solid remits of exploration consisted precisely in the search and quest for our origins? I agree that this new hypothesis is a little excessive but Christopher Columbus and Ferdinand Magellan certainly did not agree when they took on the oceans of the planet; nor did the Aéropostale (the famous French airmail service) pilots who entered the doldrums of the South Atlantic or the high valleys of the Andes. And yet, wrote Antoine de Saint-Exupéry, when the great Jean Mermoz crossed the Andes putting his life at risk, when he carried love letters and letters from merchants putting his own life and that of his mechanic at risk, he well and truly 'discovered his true identity'. What if the crisis faced by space exploration today did not so much involve technological, strategic, political and economic challenges associated with a possible return to the Moon or an arrival on Mars, or in the never clear-cut choice between man and robot, but rather in taking the search for our origins and thus our identity seriously?

It is important to agree on the meaning of the word 'origin' and to distinguish it from 'beginning'. Scientists who study the history of the universe or that of terrestrial life, ourselves included, when we ask ourselves about our own existence, can all easily feel like we are hitting insurmountable limits as soon as we try to find out how it all began. In his autobiography, Charles Darwin recognised: 'The mystery of the beginning of all things is insoluble by us; and I for one must be content to remain an agnostic.' Even if we undertake the most daring introspective explorations, we will always arrive too late to be contemporaries of our personal or cosmic beginnings. Even if we follow the winding paths of the imagination, examine the looks of the dying, weigh upon hearts or devise incredible cosmological formulae, we will still be ahead of our ends, our deaths, our passings, those of our people and those of our civilisations or that of our world. Beginnings and ends, which are only reflections of the former, remain forever unknown and, as such, continue therefore to fascinate and to tempt explorers, eager to push the boundaries already reached further and further away. And yet, there we only find volatile and ephemeral states, which largely depend on the definitions that we give to the methods of existence they surround. The notion of 'origin' seems more risky yet more appealing at the same time.

Talking about the origins of a person involves clearly identifying their date and place of birth, looking at the information on their identity papers, at their family and country of origin, at their culture and social environment—that is anything that may constitute or influence, at one time or another, their behaviour, their choices, their non-acceptances, etc. The idea of 'origin' is to understand from the point of view of originality, the singular nature of a person in the present. *Hic et nunc*, here and now.

Failing to distinguish 'origin' from 'beginning' is taking the risk of understanding reality to be a tireless repetition, an inevitable imitation of an archetype resulting from the past and a single beginning. Two beliefs, which possess a religious dimension, can be associated with a confusion of this kind: all forms of innovation are potentially harmful; human life is understood to be a fall from the divine sphere, from a primordial and perfect state. When confused with 'beginning', 'origin' loses its capacity to generate vitality. Looking in depth at our origin must not therefore only be about looking at dried bones or ancient fossils. It must be about asking what makes us original, look-

ing at our individual natures and understanding the simple fact that
. . . we are the contemporaries of our origin! What initially appeared
to be a statement of the obvious turns out to be extremely effective,
taking us back to the root, the principle and what we commonly refer
to as the origin, all questions and all answers, any curiosity and any
desire for knowledge, any propensity to explore.

The past and the future belong in some way to the field of beginnings. Knowledge of the past, whether astronomical and paleonto-
logical discoveries, is always simplified with regard to the immensity
of history, always stamped with a hypothetical character. The future
remains inaccessible to our knowledge, to our dreams and to our
beliefs, to our prayers, to our religious beliefs and to our faith, at least
in its concrete realisation.

However, if we consider origin in its contemporaneity, we must
recognise our existence and that of those around us as a fact that is
imposed on or is proposed to us, through the objectivity of things, in
their immediacy, even delayed like that of the stars of the universe. All
existence becomes an event offered to the conscience and knowledge
of humans, directly or using the instruments they have designed. Origin opens itself to exploration.

What if these explorers had all secretly chosen to prefer the origin
to the start and the end, leaving open therefore the question of the
sense of reality they discover?

7

Space at all Costs?

Yet, how many explorers and how many academics have paid with their lives to satisfy their thirst for knowledge, their thirst and that of the human species? What limits should be imposed? At what cost should space become inaccessible?

'In the kingdom of ends everything has either value or dignity. What has a price can be replaced by something else as its equivalent; on the other hand, what is raised above all price and therefore admits of no equivalent has a dignity.' Often asked about the cost of space activities—a cost often deemed by the public to be astronomical, exorbitant or staggering—I believe it is relevant and useful to refer to the distinction made by Emmanuel Kant in his work *Fundamental Principles of the Metaphysic of Morals*.

It is commonplace to represent space budgets committed to by Governments, quoting figures and ratios, comparing budget lines and public programmes: the cost of a now-retired American space shuttle mission with the cost to build a hospital in Paris; the cost to launch the European rocket Ariane 5 with the price of the French fighter plane Rafale; the cost of civil space programmes and the military space programmes; etc. And it seems inappropriate to give the budgets of private companies, including those allocated to broadcasting top sporting events or lottery sales.

These examples adequately demonstrate that space activities are among the largest investment and financing programmes committed to and undertaken by Governments. No more, no less. Why are they therefore considered to be too costly? Why consider that 'with such amounts we would be better off researching cures for cancer, eliminating hunger, or investing in other worthy causes'? Should we be seeing the effects of the clear process of trivialisation that is affecting the most useful space tools? Inhabitants of Earth so quickly got used to being flown over by communication, observation and positioning satellites that they forgot about them, surprised that they are still being designed and constructed, launched into orbit and maintained to provide multiple remote services which they are happy to receive day after day, such as weather forecasts to help navigation and observation and surveillance international communication networks. These discreet satellite systems that are invisible to the naked eye have both a cost and a price, in the Kantian sense of the term. They have a price that can or should be compared and assessed with respect to the accomplishment of a mission, the obtaining of a result, the realisation of an end, and also with respect to equivalent earth services when these exist. Does more information need to be transmitted? Today, it is necessary to compare the benefits and costs of a satellite and fibre optics, according to accessibility for the populations concerned. In this case, this comparison leads us to combine two techniques. It would be possible to proceed in the same way for all techniques and space operations that belong to or constitute greater Earth and that offer us their services. Space has a price yet is without a doubt not cost-prohibitive.

Space also develops the field of activity of exploration, which involves space telescopes, planetary probes and, to a lesser extent, manned flights. Whether the cost of these missions can be deemed modest (a space telescope that detects exoplanets cost circa 200 million euros) or whether it is more expensive (an automatic Mars mission today costs several billion dollars), it is important to realize that space has no equivalent in this sense and, therefore, must be measured and assessed not based on cost, but in other ways. It should also be considered from the point of view of Kant's second notion of dignity. This position is not excessive or misplaced. To increase human knowledge of the universe, life and their origins, to face worlds of which we were previously unaware, to risk challenging ideas, theories and certainties: what would have become of our species if, since its birth, it had not stayed true to its natural curiosity and perhaps innate penchant for exploration? What would become of our species if we decided to no longer be driven, influenced or inspired by exploration? The question is important enough to challenge the roots, conditions and purpose of exploration. I would just like to raise the issue of dangers encountered, risks accepted and to expand on the notion of dignity.

Astronauts continue to inspire dreams. Through these dreams and by the flag they bear on their space suit, according to a well-known identification phenomenon, they are citizens from an entire nation and inhabitants of an entire planet, taking part in the adventure of manned flights and who, indirectly, follow the route of the stars. Astronauts earned the title and role of envoys of humanity. However, this dream, especially following the Challenger and Columbia shuttle disasters, must be confronted by reality. Since the first manned flights, almost five hundred men and women have ventured into space. Of those, twenty two were killed in accidents during a mission or during their training—a number which is far from negligible. It is not surprising if public opinion questions the appropriateness of pursuing such a risky and humanly expensive venture, adventure, as that of manned space flights. Does space exploration exceed the limits of what is acceptable?

'Humankind', states François Ewald, 'is an animal doomed to risk.' Is risk specific to humans? Does the simple fact of existing not lie at the root of most risks faced by all living beings? There are natural risks against which scientific progress can only ever offer partial or temporary protection, risks for which only destiny, biological chance, fatality or God may be summoned. However, technological, industrial and military accidents and professional misconduct belong to the cultural sphere and were created by man 'due to his inability to master all the elements of the systems he has designed, in his haste to apply on a large scale solutions or products that have not been tried and tested, through his failure, recklessness, violence or unreasonableness, etc' (Jean-Jacques Salomon). Effectively, if we look at it this way and consider the role played and the place occupied by technology, the human being is an animal singularly doomed to risk.

We have a very strange relationship today with the notion of risk. On the one hand, we demand that public and government institutions, at least those in western societies, manage and increasingly control risks to which individuals and groups are constantly and inevitably exposed. On the other hand, we are ready to choose attitudes, to undertake high-risk activities—in these I include the taking of drugs and the practice of extreme sports. This strange relation may become paradoxical or a source of conflict when an individual who

consciously engages in a high-risk activity expects society to help
them out of danger. Many debates are taking place around Europe
concerning the practice of mountain sports and the financing of res-
cue teams which these extreme sportsmen and women may need to
call upon. This sensitive question is linked to the relationship between
the private sphere and the public sphere; a relationship that cannot be
fixed once and for all; a relationship that legislation can only partially
fix. This is particularly true when it comes to landing on new planets,
acquiring new knowledge and exploring.

The risk, this 'fear of evil' referred by scholars and philosophers
in the seventeenth who introduced the notion, is proportional to the
seriousness of a danger and to its possibility of occurrence. Any decision made in this respect is based both on the objective probability of
occurrence and the subjective inappropriateness of surrendering to
it. Inevitably associated with the design of a world entirely controlled
by the Gods or by destiny, associated with the modern introduction
of statistical calculations, the notion of risk has been or has been seen
to be ignored by engineers for a long time. The zero-risk cult, from
the absence of failure long ago clamped down on behalf of the single
objective to be achieved, the single success to be planned. Thinking
of failure seemed, to say the least, inconvenient, a waste of time and
energy, a weakening of the ideology of progress and success. After
release of the *Apollo XIII* film, the words commonly, and probably
falsely, attributed to Eugene F Kranz, the famous director of flights at
NASA, '*Failure is not an option*', became the slogan for a generation
of engineers.

Times have changed. The people responsible today for developing and operating space launch vehicles and vessels now recognise
that whilst a certain level of risk is inevitable, risks can be managed
and reduced. They must abandon simplistic and determinist explanations of phenomena and the ideal of unlimited rational knowledge,
to focus on an understanding and an apprehension of reality under
the mode of complexity and the tangle of factors, most often using
statistical calculations. It is not a question of losing the voluntarist
character that is true of the profession of engineer. It is important to
consider the increasing complexity of systems implemented and of
sciences and techniques in general. It is also important to make sense
of growth, at the interface, the intricacy of equipment, or organisa-
tional matters and of humans, in a way that the supposed skills and

responsibilities seem to disappear. This is a very strange alchemy as now an object, a technology, an operation and a treatment are often declared certain, especially when their associated risks are known and deemed acceptable by their users! This is why we fear aircraft accidents more than we fear road accidents. Is it enough to invoke the enlightened or informed consent to ward off fear of evil? Not necessarily, as knowledge can lead to other fears. A dilemma without an apparent solution or outcome or a dilemma that is associated with the single truly responsible situation, a dilemma based on absolute conviction and the double stubborn refusal of fatality and the scandalous 'After me, the deluge!' This courage and this sense of price to be paid and of dignity to be defended are necessary, when, tomorrow, astronauts could encounter new dangers, new risks, new sacrifices. Will they be ready to accept them? Will we be ready?

8
Heroism or Suicide?

Mars One, the commercial project launched by the Dutch engineer Bas Lansdorp to place a colony on Mars by 2024, made headlines and generated a range of different reactions. Lansdorp claims that a manned mission to Mars is perfectly feasible, although space agencies such as NASA are constantly putting off the idea, for both financial and technical reasons. The solutions suggested by Lansdorp are highly debatable and have been widely challenged and criticized, especially the way the project would be financed (by selling the media rights to cover the expedition, rather like a reality show) and the fate of the travellers.

The high media profile achieved by *Mars One* is perhaps due to the ambiguity of the programme's name, which can be interpreted in at least two ways. Should we take it at face value, as suggested by the website that shares its name? In other words, a project to send the first manned mission to the Red Planet, not just an exploratory mission but a project to set up a living area for a team of four humans, to be joined two years later by a second team of four sent from the Earth. In this case, Mars One would just mean the first colony on Mars. Or could it be a case of Mars One Way, with no return ticket? Because the project's organisers have made no secret about it: there will be no way of getting back to Earth, even for a well-deserved holiday.

Is it this second interpretation that has aroused all the media interest in the project? It would be the first time humans had been offered the opportunity to leave Earth and never to return: suicide, or sacrifice? No-one had ever asked such a question before. When President Kennedy sent his country off on the Race to the Moon at the beginning of the 1960s, he set out the conditions very clearly: that the first American to set foot on the Moon must also return to Earth safe and sound. Those who defend Mars One can quite reasonably counter this objection, or rather this reminder of the spirit in which the first space heroes set forth, by explaining that their project is to colonise Mars and not to explore it; in other words, the fact that the crews will never return to Earth is not the goal, nor the means, nor even a consequence, but is integral to the notion of colonisation. Suppose that we accept this argument, while regretting the possible—and perhaps intentional—ambiguity, which arouses so much public interest, though perhaps on false pretences.

Mars One thus proposes to establish the first human settlement on the Red Planet. The history of our species certainly includes similar examples of colonisation that occurred without prior exploration, however brief, of what was previously unknown territory; just as Mars remains today, despite the success of several unmanned missions, to and on the surface of Mars or in orbit around it. But do we know enough? By which I mean, do we have enough data to install a human colony which is anything more than just a kind of submarine or a space station that has come to rest on the surface of Mars; and also enough to ensure that the many operations planned by the Mars One project do not involve catastrophic and irreversible consequences for the Martian environment? The draconian sterilisation measures applied to the robots placed on the Red Planet suggest the opposite. Before any human takes a first step on the surface of Mars, mankind should probably take a much bigger leap in acquiring technical and scientific knowledge necessary for planetary protection. And we should not forget that, under space law, Mars is part of the common heritage of mankind; in other words, no one can do just whatever they like there: each of us may have rights regarding this planet, but also obligations.

Now let us consider the selection process. Proposing the Mars One programme, or getting oneself selected for it, is obviously a wonderful way of being talked about. This is what many of the project's detractors have thought, even adding that it was entirely risk-free since, according to most specialists, today's technologies that the founders of Mars One propose to use are not up to the task, and are unlikely to be so within the timetable put forward. We only need to look at the proposed suborbital space flights for tourists: the prototype, Space Ship One, made its maiden flight in 2004, but the first commercial flight is currently planned for 2015 at the earliest. Even if Mars One could realistically use off-the-shelf technologies that industry has already developed, it seems very likely that there would be delays. But why puncture the dreams of these potential colonists of Mars. Do we not all need to dream, to have something to enthuse about? Since the Copernican revolution, space no longer attracts only the soulful and the mystical; it has been opened up to the scrutiny of the curious and to explorers. It inspired the heroes of America's space programme in the twentieth Century who had 'the right stuff'; so why should it not continue to do so in the twenty-first, even if those who are selected should in fact, contrary to their expectations or what they have been told, be able to return to Earth? What would mankind become if it ever started to despise dreams, and ceased its eternal march to the stars? We build and strive for much more dangerous castles in Spain than those promised by Mars One.

One last point. Forget 'Mars One'; think '*Press one to vote for X, press two to vote for Y*'. The reality-show aspect is intended not just to finance the project but to enable the public to be in on the final selection of the first crew to leave Earth for Mars. So be it. But I think everyone of us should try and think how we would feel if the crew we had voted for were to perish during the journey to Mars or in the first few weeks after landing. Are *you* ready to bear the responsibility for their deaths? Who has the cold determination of a von Braun or of his Russian equivalent Korolev, when the moment comes to choose a twenty-first Century Gagarin, Armstrong or Aldrin? Even when it seems like a game, space remains terribly, and tragically, human.

9
Does the 'Envoy of Humankind' Concept Have a Future?

London, Moscow, Washington, 27 January 1967: the signing of the UN *Outer Space Treaty*. In Article 5, we find the following: 'States Parties to the Treaty shall regard astronauts as envoys of mankind in outer space and shall render to them all possible assistance in the event of accident, distress, or emergency landing on the territory of another State Party or on the high seas.'[1]

Envoy. The etymology is highly revealing. The Latin term introduces de facto the idea of a way (*in via*) and of making one's way (*inviare*); in other words, space and time, beginning and end, a journey and its ultimate destination, an outward and a return journey. Highly revealing indeed, because there could be no dispatch of envoys without all this and so many other things that may well belong, if not to its defining characteristics, at least to the identity of the human species. Is the bee which goes in search of pollen an envoy dispatched by its hive? Is the bird hunting for insects an envoy dispatched by the young it has to feed? And are they consciously fulfilling that role? The answer is not our basic concern here; but when it comes to human beings going into space, the United Nations has seen fit to make them envoys, and not only that but envoys of mankind. For the first time in the history of our species, an individual was being given—and astronauts are to this day given—the status of representing humanity at large. This is one way of implementing the provision that the 1967 Treaty introduced in Article 1: 'The exploration and use of outer space, including the Moon and other celestial bodies, shall be carried

---

1. *Treaty on Principles Governing the Activities of States in the Exploration and Use of Outer Space, including the Moon and Other Celestial Bodies (OST).* Article 5, first paragraph.

out for the benefit and in the interests of all countries, irrespective of their degree of economic or scientific development, and shall be the province of all mankind.'[2]

With specific reference to the French text of the *Outer Space Treaty*, what is to be understood by the term *'apanage'* ('province' in English)? In France, at the time when the country was still ruled by the monarchy, *apanage* referred to that portion of the royal estate that was given to younger members of the royal family as compensation for being denied the aspiration of being heir to the throne; since then, the term has come to designate more broadly a possession or an inheritance, with the added connotation of exclusivity. The choice of the term *apanage* by the drafters of the French version of the text is an interesting one. For one thing, the French term *apanage* strikes a tone neither of dominance (since we are not masters of the universe) nor of submission (and neither are we its inheritors). Moreover, that inheritance is not primarily a territory as such, but rather a mission: that of exploiting and using outer space in the interests of present and future generations. Declared envoys of mankind, the prime mission assigned to astronauts is accordingly to safeguard this inheritance, for the good and in the interest of not only the space powers but of humanity at large. Since 2007 marks the fortieth anniversary of the drafting and signing of this basic text of international space law, a fundamental question arises: have our astronauts and those in charge of space activities actually managed to put these provisions into practice? Are astronauts actually regarded as envoys of the entire human race?

To begin with, we have to accept reality: overcoming nationalistic impulses and habits is clearly no easy matter! We have only to consider the insignia borne by these so-called envoys of mankind. Their nationality is always on full view. Not to mention the six US flags that have come to adorn the surface of the Moon. Astronauts are indeed envoys dispatched by humankind; but dispatched there thanks to explicitly identified national taxpayers' money. Could we imagine Neil Armstrong in front of the Lunar Excursion Module planting the Earth's or simply the United Nations' flag, especially given that his extraordinary venture into space owed so much to the Cold War and to the colossal financial effort made by the United States? And in any

---

2.   *OST.* Article I, first paragraph.

case, how can we demand such high standards of non-partisanship in an area in which the strategic and economic stakes are so high . . . and when the Olympic Games themselves are all about confrontation, albeit of a sporting nature, between the nations of the Earth, with flags and national anthems taking centre stage?

Rather than simply lament this persistence of nationalism and specific interests, which has extended to the realm of space itself, it would be more productive to point out that the missions conducted by these astronauts and their accounts of their experience of space have truly helped to boost our collective self-awareness as a species. To a certain extent, Hannah Arendt was wrong when she wrote in her book *The Human Condition* that the launching of the first Sputnik was an event second in importance to no other, not even to the splitting of the atom; and that event marked the fulfilment of the prophecy uttered by Russian scientist Konstantin Tsiolkovski, engraved on his tombstone, that mankind would not forever remain on Earth. Rather than refer to the Earth as the cradle of humanity, astronauts nowadays much more often refer to it as a spaceship in our charge. Serge Brunier is right to point out that the glossiest publications presenting the Earth as photographed from space by astronauts and satellites have never exceeded a print-run of around 100,000; whereas in contrast, *Earth From Above* (published by Abrams), featuring pictures of our planet taken by photographer Yann Arthus-Bertrand from a helicopter flying at an altitude of under 1000 metres, has sold more than three million copies in over twenty languages.[3] But, despite having no glossy book of photographs to their name, what can be said about the views of the Earth brought back by the Apollo mission astronauts except that nearly forty years on they are as popular as ever and their influence on contemporary environmental awareness is undeniable.[4]

No doubt up until now it is Mother Earth that has carried humankind, and indeed continues to do so, like a mother carrying her offspring. But we have surely now reached the stage where it is time for us to take our turn to do the carrying. And it may well be that never before have we as a species had such a keen sense that our destiny,

---

3. Cf Serge Brunier, *Impasse de l'espace. À quoi servent les astronautes?* (Paris: Seuil, 2006).
4. Cf Denis Cosgrove, *Apollo's Eye. A Cartographic Genealogy of the Earth in the Western Imagination* (Baltimore & London: Johns Hopkins University Press, 2001).

and that of this planet that is our home in space, are so closely inter-twined.

There is no doubt that the change in the tasks handled by astro-nauts has tended to take some of the shine off their status as envoys of mankind. As early as 1982, one expert observed, regarding these tasks, that the conquest of space had essentially become a battle on three fronts—commercial, political and strategic—with the explorer replaced by the soldier, representative of business, and investor. Gabriel Lafferranderie adds that:

Astronauts are no longer just pilots, but also scientists, astrono-mers, doctors, engineers and journalists. They eventually will become gardeners and miners too. The experiments they conduct are on behalf of industry. These 'multi-tasking' astronauts live in a confined space and are watched and listened in on permanently by ground control, which sends up their working instructions, awakens them or requests that they sleep, exercise or undergo medical experiments and perform movements in space that seem completely unnatural.[5]

Multi-tasking astronaut might seem a surprising term to use. Nev-ertheless, it is an accurate one and begs the question: what similarity can there be between an envoy of mankind bearing the stamp of the UN and a multi-tasking astronaut, or even a space tourist, who is after all sent up only by him or herself, a few sponsors or a TV pro-duction company?

European astronauts were aware of all these issues when they drew up their Charter in summer 2001. The vision, mission and values they defined were directly inspired by the founding principles of the space venture, as set out in the texts already mentioned above. Conscious of being part of a cultural and historical process and concerned with working for the good of humankind, they took care to emphasise the risks associated with the space endeavour, and both the daring and wisdom it required.[6] The Charter has served as a basis for reflection on the ethical implications of possible (partial) commercialisation of the activities carried out by European astronauts. Such commerciali-sation could call into question the status they have enjoyed thus far, as well as the image they wish to portray to the public and the image

---

5. Gabriel Lafferranderie, 'Espace juridique et juridiction de l'espace', in *L'Homme dans l'espace*, edited by Alain Esterle (Paris: PUF, 1993), 255.

6. The initials of the five values promoted by the charter form the word *SPACE*: *Sapientia, Populus, Audacia, Cultura* and *Exploratio*.

expected of them by that same public. It is not enough to set up an ethics committee tasked with assessing the quality and appropriateness of proposals for sponsorship and remuneration. One needs also to ask at which point adherence to a status grounded in law becomes a conservative reaction and acts as a brake on imagining and foreseeing the future.

10
Avatars

Versus: this is the word that comes to mind when we start to think about possible ways of implementing the human penchant for exploration. *Versus*: because, all too often, these ways are presented as competitive, as exclusive with regard to each other. Human *versus* animal, human *versus* robot: in the same way than legal cases in the United States are labelled using this Latin term or one of its abbreviated forms (*v* or *v*, *vs* or *vs*). The alternative often concerns the conflicting appearance of a face-to-face meeting, or an altercation between the supporters of each of these three possibilities.

Claude Bernard was the first to use the animal model in biology and invented the experimental method in medicine. Man and animals in fact share a number of physiological processes. The space domain is not exempt from this practice: from the beginnings of space adventure, animal took the place before man. At the time of the first space flights, the dog Laika went before the Russian Yuri Gagarin and the chimpanzees Ham and Enos went before the American Alan Shepard. This was to gain a better understanding of any physiological changes and pathological consequences that may affect the pulse, breathing, arterial and venous pressure, under the effect of acceleration, altitude and microgravity. More than thirty-five species of animal have been sent into space over a period of fifty years.

There are many arguments to support the use of the animal model. Those of an ethical nature primarily consider the moral limits in terms of investigation and exploration on a human being. As current legislation stands, biopsies, functional destructions, the implementation of sensors in the inner structure of the body are only possible with animals. In addition, the potential for experimentation on humans is restricted due to the absence of the possibility of any therapeutics in the event of an incident or accident. According to regulations in force, *in vivo* experiments on animals is itself limited to strict necessity. Any which way, there would seem to be three imperative rules: experimental protocol is compatible with health requirements, the procedure used causes as little trauma as possible, clinical examination is performed on a daily basis.

Scientific arguments more specifically concern experimentation conditions and the population of models and their environment must be as homogenous as possible. Astronauts are themselves from a heterogeneous population (physical and nutritional conditions, psychological condition, level of training, degree of familiarity with the space environment, workload) and it may therefore seem difficult to compare the quality of the batch (if I may so call it) they could form, with that of a laboratory animal population specifically selected for their homogeneity and placed in permanently controlled conditions.

Arguments of an operational or methodological nature are primarily related to the only partial availability of crew members, with respect to experiments, notwithstanding their small numbers; tests cannot be multiplied either before or after the flight. On the other hand, tests can be multiplied with the animal model, including on experimental subjects similar to those which took part in the flight. However, it is important to consider the drawbacks associated with the use of animal models in space: the possible unpleasantness in the event of the inadequate sealing of maintenance modules, for flights on which both humans and animals are aboard; the constraints associated with feeding; the retrieval of urine and faeces; the sensitivity to stress.

There have been several controversies regarding the use of animals in space biology programmes. The usual arguments are apparently no longer adequate in responding to uneasiness or doubt among the general public. At the beginning of 1999, the *Space News* review published an article by AR Hogan and ND Barnard, promoting a greater respect for our 'terrestrial relatives', the expression now used to designate animal species. Both authors refer to the unpleasant aftermath of the BION 11 mission, in 1996, and the death of the monkey Multik, and the surprises of the Neurolab mission, in 1998, involving the death of many of the rats on board. They focus on the regrets of Oleg Gazenko, the former head of Soviet animal testing in space programmes. Several weeks later, Gazenko specified that, without going back on his regrets ('We did not learn enough from the mission to justify the death of the dog Laika'), he considered as nothing less than decisive the contribution made by animal experimentation to the progress of physiological knowledge, both demanded and authorised by the development of space technologies. Space does not therefore escape the paradox often apparent in our western societies: animal testing seems to provoke more doubt, reluctance and opposition than tests conducted on humans! Can the reason for this be solely put down to the fact that humans possess a conscience, unlike our terrestrial companions? Can the debate be reduced to the question that appeared several years ago on a web page: 'Is planting a flag on the Moon or on Mars worth the life of a dog or a monkey?' In any event, the use of animals for experiments in space was the subject of intense discussions within ESA. In Spring 1999, an agreement authorizing the use of laboratory animals in space was concluded between all partners, excluding Sweden. However, several voices spoke out in favour of the establishment of a committee responsible for drawing up ethical rules pertaining to animal experimentation.

Animals in space are given either the status of guinea pig, of substitute or of emissary.

In theory, the status of guinea pig does not require any specific comments: firstly, it is comparable with that of animals used by laboratories on Earth and globally raises the same questions, primarily that of the necessity for their use. However, one of the questions and the only one that may be worth holding on to as being specific to space concerns the possibility of managing the well-being of these animals. It would be difficult to send work inspectors or health inspectors to check the scientific practices applied in space stations and spacecraft, especially for unforeseen and discrete inspections. Incidentally this comment also concerns the application of laws and regulations regarding experiments on humans.

The status of substitute is more specific to space. By this expression, I mean animals which take the place of humans on board space vehicles to test their reliability prior to the boarding of humans. The ex-USSR, the USA and also China have sent animals into space in order to prepare for their future manned missions. It is not strictly speaking a question of 'backup' (an expression used moreover in the domain of manned flights), as these are intended to replace a faltering human being and not to precede him. In looking at the photographs of Laika or Ham, it is however obvious that the animal literally takes the place of the cosmonaut or the astronaut. Seeing the helmets, suits and harnesses worn by these animals, how is it possible not to talk about simulacrum? Or better still an emissary?

'Soviet' dogs and 'American' monkeys that preceded the first human space flights were emissaries. The term is not too strong: there is something about space that continues to elude us, that belongs to a world which is not our own, and that both attracts and frightens us. To come into contact with it and travel into it cannot be done without taking great care, and therein lies clearly the nature of the sacrifice involved. In other words, what is unknown or strange to us, all that which much of space still represents for us, must be tamed, while at the same time we ourselves must be tamed by space. This requires time and maybe also an emissary, a scapegoat to use the language of sacrifice, sent out ahead and in place of others. That was indeed the role (or mission) formerly entrusted to animals in space and might again be so in the future. Between now and the arrival of the first tourist on Mars, time and technological advances will be needed, as well as numerous exploration missions, with living beings onboard, non-human to start with, and then humans themselves. New envoys, in other words. But are we ready to accept animals as the emissaries of mankind? It is worth going over the pages in writer-journalist Tom Wolfe's work *The Right Stuff* devoted to how American test pilots viewed their colleagues selected for the Mercury space programme. Charles Elwood 'Chuck' Yeager, the first pilot to travel faster than sound who did not belong to the group of the first American astronauts, explained to a group of journalists:

> I've been a pilot all my life, and there won't be any flying to do in Project Mercury.
> -*No flying?* - [. . .]
> The thing was, he said, the Mercury system was completely automated. Once they put you in the capsule, that was the last you got to say about the subject.
> - *Whuh!*
> - Well, said Yeager, a monkey's gonna make the first flight.
> - *A monkey?*

The reporters were shocked. It happened to be true that the plans called for sending up chimpanzees in both suborbital and orbital flights, identical to the flights the astronauts would make, before risking the men. But just to say it like that . . .! Was this national heresy? What the hell was it?[1]

Yeager talks about the heresy of reducing the pilot to the rank of animal. On the contrary, French philosopher Florence Burgat talks about denial: the staging and photographs of space animals try, conversely, to eliminate the animality of the space dog or chimpanzee, to elevate them to practically human level.[2] Do we have to choose between these two extreme positions? I do not think so and prefer to retain the idea of emissary, of being sent in the place of; while this also means accepting and deciding to sacrifice emissaries. Buzz Aldrin, the American astronaut who took part in the Gemini and Apollo programmes reminded us in June 1998 that he and his colleagues appreciated 'the enormous debt we owe the space chimpanzees'. And he added: 'Now it is time to repay this debt by giving these veterans the peaceful and permanent retirement they deserve.' The astronaut's project at that time was to raise the money required to place these chimpanzees in a Texan park, under the protection of a specialist organisation. I have not heard whether this project has been completed, but it at least demonstrates how it is possible, even before drawing up animal rights (those of space or Earth), to consider the responsibilities of humans towards them.

If the requirement to use animals in space arises again, it will necessarily be to explore new domains and to cross new frontiers. This applies in particular to long-term missions. Which technologies need to be established to successfully complete missions lasting dozens of months, in zero gravity and subject to strong cosmic radiation? In fact, I am currently unaware of any discussions regarding animals going on board prior to the first missions with humans to and on Mars. However, if it were to become or when it does become the case, I believe that we will be in a position to better consider the meaning given to this venture, which cannot find its justification in the free and informed consent of a team of astronauts and in the genius of scientists and engineers alone. In other words, the use of animals,

1. Tom Wolfe, *The Right Stuff* (New York: Farrar-Straus-Giroux, 1983), 127.
2. Florence Burgat, *Animal, mon prochain*. (Paris: Editions Odile Jacob, 1997), 149.

whether they are considered as emissaries or substitutes for humans, necessarily raises a question and requires an ethical decision, which may be dealt with by the space community itself: the decision to send a living species into space is based, I believe, on the scope of skill and responsibility of an entire society and not a group of experts. If the scientific and technological stakes of such an operation are real and often immediate, the debt with regards to the substitutes of mankind binds us humans. This debt calls for the question concerning the finality of space exploration programmes involving human or non-human lives to be raised urgently and precisely.

While they still look like and function like a household appliance, or an interplanetary probe, robots do not *a priori* give humans any cause for concern. When an example of a new robot takes on and crosses, like Voyager 1, the confines of the solar system after a thirty-five year journey, our feeling is one of pride for a product produced by human intelligence, built by man, and travelling to places where man has not yet set foot. Philosophers did not wait for the invention of the first machines capable of self-control before asking whether, eventually, the robot would take the place of humans not only in performing the most servile tasks (which no-one would complain about) but also in performing the most noble and human of tasks, from those allying their capacity for observation and analysis, the conscience of a mission to be accomplished or of a goal to be reached, the power to choose the best way to achieve it, etc. What fell within the bounds of science fiction just a few decades ago is now a technological reality and an everyday practice.

Space certainly did not escape this development and its techniques rely heavily on the use and control of robots. In 1783, the Montgolfier brothers put three animals in one of their first aerostats to mark the beginning of the conquest of the skies, preceding Pilâtre de Rozier by a few months. In 1957, a Sputnik, an artificial travelling companion, was charged with opening the doors to space, before the dog Laika and the cosmonaut Gagarin. Since this date, the human *versus* robot dilemma has continued to occupy or be at the forefront of the space scene, neither camp finding itself deserted or short of arguments. A famous debate on the subject took place in France during which a member of the French Academy of Sciences explained how, due to the progress of artificial intelligence technologies, robots could soon entirely replace humans on all space missions. A cosmonaut replied:

'Will academics also be replaced by robots on that day?' Twenty-five years before, one of the Apollo astronauts said it was a shame that a poet could not take his place to express the extraordinary beauty of the Moon . . . whilst recognising that the success of the mission would without doubt have been compromised!

Regardless of progress made by scientists and engineers since the end of the 1960s, none of those involved dares to promise us the development of a robot poet. However, what they do give the astronauts, for example on board the international space station, is not far from reaching the achievements of their ancestor HAL imagined by Arthur C. Clarke in *2001, Space Odyssey*. If we also take the example of *drones* that not only survey cities and battle fields but carry weapons, that of *rovers* which travel across the surface of Mars and manage their own missions, and finally that of descendants of the Sputnik, which tirelessly continue to circumnavigate the Earth, how is it possible to refrain from asking if it is still necessary for a pilot to be in an aircraft, a human to be in a space vessel or on board an exploration vehicle? Is it not sufficient that they are comfortably seated in front of their computer screen with a joystick in hand? Even at the risk of losing a little poetry and a lot of sense of responsibility.[3]

I do not believe that we need to continue to pit man against robot. This attitude can only lead to broken promises and harsh critique. However, it is advisable to raise the serious issue of the consequences of a growing cooperation between humans and their increasingly sophisticated machines. Once again, we turn to works of science fiction to support these claims. The film *Avatar*, directed by James Cameron in 2009, deals with the colonisation and conquest of 'new worlds', with the abusive use of resources on which living species and environments depend, with the conflict between cultures and civilisations and with conflicts between military operations and scientific projects. But the title of the film and the imaginary reality it portrays is important. In the words of Cameron himself, *Avatar* is the ability to 'inject a human's intelligence into a remotely located body'. Using the technology of the avatar and this intervention derived from genetic engineering, neurosciences, other sciences and science fiction, Cameron offers a solution to all those whose passion for exploration (or

---

3. Cf Jacques Arnould, *La Terre d'un clic. Du bon usage des satellites* (Paris: Odile Jacob, 2010).

intention of conquering) grapples with the limits of the human condition and its technologies, with the inhospitable nature of terrestrial environments and extraterrestrial environments, with the constraints of space and time. The Canadian director had to wait several years for cinematographic techniques to be advanced enough to be able to screen the Pandora landscapes and the bewitching Neytiri (the virtual avatar of a real actress). Is it really wrong to dream or to think that, in the not too distant future, human avatars, no longer virtual but real, could begin to conquer near or far space, free of the constraints of human skin?

The development of such avatars would far surpass the current abilities of the most advanced, agile and autonomous robot explorers. Aside from the scientific and technical challenges, it would raise a wealth of philosophical and ethical questions. When 'I become another', it is important to ask 'who am I?' Current research in the field of artificial intelligence is already raising questions about the place and importance to be granted to biological support in the possession and exercise of a conscience, feelings, etc. Walter Pons' question 'Is the sky open to us?' and the invitation to know oneself that he extended to his readers in the 1960s, are relevant to the, still fictional, prospect of explorer avatars.

In addition to the questions raised by the screening of *Avatar*, I remind you of perhaps more evident questions relating to the ethics of exploration (and perhaps innovation also). To what extent will scientific, technical, political and economic capacity, in addition to the courage to undertake to and succeed in exploring other worlds than our own give us power over them, over what we see as potential resources, over species and environments, intelligent life forms and cultures that we might discover there? What rights would we have over them and what would we owe to them? Astrobiological researchers and lunar prospectors must not ignore such questions even though they do not yet anticipate an imminent meeting with Pandora's Navis or the discovery of *unobtanium*, the rare mineral coveted by humans in Cameron's film. Other films such as *Encounters of the Third Kind*, directed by Steven Spielberg in 1977, taught us that meeting others, a stranger, an alien, is always overwhelming and can lead to our conversion to another culture or another identity. Such are the dangers and opportunities that real exploration offers.

11
A Virtuous Circle

The Right Stuff. Because space exploration requires humans they accept to face extraordinary risks, because it submits them to a paradoxical situation of pride, due to the amount of their scientific and technical knowledge, and modesty, in comparison to the dimensions of the universe, this enterprise concerns, affects and shakes what Tom Wolfe judiciously chose to call their inner *stuff,* meaning their fundamental essence, the very core of their being. This does not simply mean the basic raw material of human nature, by which I mean, for example, their genetic make-up, which, as we know, influences the way every human being behaves when faced with the chance events that life can throw up, by coming to terms with risk, avoiding danger or, on the contrary, seeking it out. It also means the more sophisticated, constructed side, the fruit of each person's own will and efforts, in short, what philosophers agree to call virtues.

It is very difficult to make do with only one definition of virtue. Readers of Stendhal may remember the definition he proposed, or rather the example he gave: 'Some virtuous and tender women have almost no idea of physical pleasure; they have rarely experienced it, if I may put it thus' (*De l'Amour,* 1822). The type of virtue of which Stendhal wrote in this instance is the kind that avoids exposure to the slightest risk; it is the adornment of those whom no one would ever confuse with women (or indeed men) 'of little worth'. However, to find a more appropriate definition of virtue for the theme of exploration we are discussing, we would do better to go back to the usual etymology of the word, which refers to virility, valour, courage and moral steadfastness, and without suggesting that these are uniquely masculine qualities. When philosophers try to define the concept of virtue, they prefer to speak of personal excellence, a lifestyle acquired by constant care and exercise, so as to be able to mobilise one's natural faculties, one's emotional and mental strength, for moral purposes. It is traditionally associated with *habitus*, a term that we should resist the temptation to translate as 'habit' because virtue, as Kant reminds us, is not fostered by repetition, but on the contrary requires constant renewal and revitalisation from its source. Again, according to the philosophical tradition, the search for good cannot rely on chance alone, nor on custom or social pressure, but only on individual freedom based on strength of character bolstered by a type of know-how. In short, virtuous people are those who lead a responsible existence, taking constant care of their own development and that of others. Philosophy offers a far more worthwhile definition of virtue than that practised by Stendhal's 'virtuous' women, or the one so often devalued or even mocked in our own time. Virtue should be understood, lived and sought as the breath and salt of life, the perfume and charm of our existence, something that certainly contributes to the right stuff, or the qualities summed up in the expression '*Noblesse oblige!*'

In the past, philosophers and theologians, moralists and thinkers tended to distinguish between the theological virtues and the cardinal virtues. The first, though still related to classical cosmology and the belief in a sort of cosmic insurance, were considered necessary to get closer to God. Defined as faith, hope and charity, they were thought to help human faculties, limited by their very nature or even tainted with evil and failure, to adapt in such a way as to accomplish God's intentions for His creatures: to enable them to participate of His divine nature. The cardinal virtues, whether justice, courage, temperance or prudence, necessarily grow out of the theological virtues, taking them further, grounding them in reality in the most day-to-day, but sometimes also the most dramatic, the most exceptional aspects of human existence. As well as discrediting the notion of cosmic insurance, modern ideas have deprived much of mankind today of any religious backing for the theological virtues: we can have faith, hope and charity without any divine inspiration. In the same way, I remain convinced that, although the cardinal virtues may have lost their theological basis, they have not lost their ability to inspire human behaviour and undertakings, especially when such undertakings lead people to test themselves to the limit, in whatever way, physically or psychologically, cultural or geographical.

The right stuff. Among the virtues required to make the human stuff able to undertake the exploration of space, two seem essential: prudence and confidence.

Prudence may no longer be a cardinal value but it still seems to be one of the primordial qualities expected or required of anyone who, for professional or recreational purposes, wishes to be subjected to or subject themselves to high-risk situations, to undertake or oversee actions with an abnormally high level of risk. There is no need to repeat here the Aristotelian debate about the theoretical or practical nature of *phronesis*. However, it might be useful to retain the idea that prudence cannot be reduced to normative knowledge, nor to the [mere] application of principles, but that it is also, and perhaps even essentially, an intelligent appreciation of situations, an ability to act for the good. Aristotle emphasizes that prudence can only be acquired over time and by taking time into account: to be prudent means learning to seize the right moment and take advantage of circumstances, having judgement and clear-sightedness, being aware of the importance of looking before leaping, of predicting and avoiding danger just as much as of dealing with the consequences. The Aristotelian notion of prudence is the virtue that enables people to combat the effects of ignorance and chance, insofar as they have the capacity, or are allowed, to do so. Perhaps the precautionary principle is the acceptable modern-day version?

When Martine Rémond-Gouilloud suggests summing up the notion of 'precaution' with the following phrase: 'If in doubt, do not abstain, but proceed as if the risk were proven',[1] she is following the same line as Aristotle, because she suggests taking preventive measures in a potentially risky situation (under the meaning defined above), even if it is impossible to know all one would like to know about a situation and to assess it perfectly. This is also the position of the French Act of 2 February 1995, on reinforcing environmental protection, known as the Barnier Act, which describes the precautionary principle as one 'according to which the lack of full scientific and technological certainty at a given time shall not be used as a reason for postponing cost-effective and proportionate measures to prevent serious and irreversible environmental degradation'. The

---

1. Martine Rémond-Gouilloud, 'Entre bêtises et precaution', in *Esprit*, 237/11, (1977): 119.

purpose of this way of applying the precautionary principle is not
to refuse to take uncertainty and risk into account with respect to
the way we act and take decisions—in short, to deny their reality in
society as we know it but on the contrary to keep these issues on the
agenda, as perfectly valid concerns; I think we can safely admit this is
a modern and acceptable interpretation of the virtue of prudence, as
taught by Aristotle. It is not entirely devoid of scepticism: how can we
ever be certain that we have correctly traced the boundary between
knowledge and ignorance, between certainty and uncertainty and,
consequently, between prevention and precaution? However, this
position leaves any definitive judgement in suspension and gives full
rein to the imagination and speculation, suspicion and doubt. Just as
in the Aristotelian tradition, it is essential and indeed fundamental
to take full account of the sequence of predictable and unpredictable
events: people who practice this sort of prudence and apply this type
of precaution take their time, avoid undue haste, striving to ensure
that their decisions and their actions will withstand the test of time,
to acquire new knowledge and, lastly, to bear in mind the lessons of
the past, but without allowing themselves to become paralysed by
indecision. Is this not one of the most reasonable, fair and effective
ways of showing our consideration here and now, as far as it is possible to do so, of the needs of future generations?

Unfortunately, as we know, this is not how the precautionary principle is usually invoked, interpreted and implemented today, especially by governments, companies, or the institutions that manage the
planet. The latter frequently manipulate it and present it as a reason
for either refusing to act or taking emergency measures. The principle
(or rather a particular interpretation of it) provides decision-makers
with an excuse for refraining from taking a position, committing
themselves to a course of action, or launching a production process
unless there is total certainty that there is no corresponding danger,
to health or the environment, for example. Hiding behind the excuse
that 'The current state of knowledge does not allow for a sufficiently
accurate prediction of long-term outcome . . .' and hypnotised by
the possibility of zero faults, zero damage, zero danger, authorities
appealing to the precautionary principle are not displaying the tra-
ditional 'virtue' of prudence. This interpretation of the precautionary
principle, and this kind of behaviour, in fact surrender almost unlimited power to scientific knowledge and technology, while forgetting

that every moment is only a point on the continuum of time, and a part of the inexorable onward march of history. They risk opening the door to authoritarian public policy, to the death of freedom, in the name of virtues that are in no way cardinal, and still less theological. How can they still claim the slightest link to the noble and reasonable virtue of prudence?

Prudence and precaution obviously have a vital role to play in the space sector. The space environment is too hostile, the factors to be controlled too numerous, the technologies too complex, for those who explore or exploit space to lower their guard; they must constantly increase the ways and means of raising the alarm and the number of backup systems, and remain aware of the danger of relying too heavily on habit or on what appears to be obvious. Looked at another way, the current state of knowledge, whether scientific or technological, far from being a reason to do nothing and take no risks, should on the contrary be one of the main reasons for facing up to such imposing difficulties and such obvious dangers. Seen in this way, the challenge is similar to that described by Bruno Latour concerning innovation:

> The innovator's dilemma is well known; when he can choose, he does not have sufficient data; once he has the data, he no longer has any choices. At the start of his project, if he knows nothing yet of the reactions of the public, financial players, suppliers, colleagues and machines that he must bring together to make his project take shape, he can however very quickly completely change the nature of his plans to adapt them to their wishes. At the end of his project, he will finally have learned everything he should have known about the resistance of materials, the reliability of components, the quality of his subcontractors, the loyalty of his bankers, the passion of his customers, but it is too late to make any further changes to his plans which are now cast in bronze.[2]

In short, it is no easy matter to choose what seems to be the most suitable moment to act, after weighing up the pros and cons and choosing the most suitable means. In the end, might not one of the principal dangers be the risk of doing either too much or too little?

---

2.   Bruno Latour, 'Comment évaluer l'innovation?', in *La Recherche*, 314 (1998): 85.

Let us take the example of risk management related to spacewalks, or the extravehicular activity, also known as EVAs, of astronauts obliged to work outside their capsules, shuttles or space stations. Among the many dangers to be taken into account to protect and preserve the lives of these spacewalkers, that of having a micrometeorite or piece of space debris, however small, striking one of them and tearing their spacesuit or shattering their helmet cannot be ruled out; at the altitudes where these objects can potentially come into contact with either spacecraft or humans, speeds can be 25,000 to 26,000 kph, and the slightest impact between them would obviously have tragic consequences. However, human beings are currently indispensable for assembling and servicing structures like the ISS. There is practically no real and effective protection equipment that the astronauts themselves could wear; the only solution is to rely on prevention by remaining sheltered by the bulk of the space station as long as possible and insofar as the work itself allows. If an accident should occur despite everything (not necessarily as a result of negligence, but simply because many objects are too small to be detected, and therefore avoided), it would be necessary to ensure that all the dangers inherent in such an operation had been identified and assessed, that every possible preventative measure had been taken, that the purpose of the mission had been clearly defined and explained, in short that the risk had been reasonably assessed. Any hesitation, any doubt regarding any of these points would immediately raise questions as to whether the virtue of prudence and the precautionary principle had been effectively applied.

Is confidence merely a degraded and more human form of the theological virtue of faith? In addition to a common etymological root, they share several features; nonetheless, many thinkers prefer not to consider confidence as a cardinal virtue at all, but merely as a feeling. No matter. Confidence, like prudence, is one of the principal virtues associated with daring enterprises, and taking calculated risks: it can include self-confidence, confidence in one's partners, colleagues and other crew members, confidence in human beings and in mankind and lastly, as a way of encompassing all these different considerations, simply – although optimism might be a better word – confidence in the onward march of time and the advance of history. This is because confidence, in order to be real, involves being lucid about the passing of time, and considering the future as being wide open with an unknown range of possibilities and probabilities, desirable or undesirable. There can be no question of claiming or insisting on confidence without genuine anticipation, without preparation and some form of adjustment, though not necessarily submission, to whatever the future holds. Being confident, or worthy of confidence, means reacting to misfortune with rational fear, and good fortune with rational hope. When confidence is lacking, our range of possibilities is narrowed: reality becomes confined and defined by the past and the present, by preconceptions and "obvious evidence" that nothing can ever change. The future is no more than an extension of the present: without confidence, there would never be anything new under the Sun. On the other hand, this leaves unanswered the question asked by Kant of anyone who places all his confidence, or asks others to place theirs, in a hazardous venture: 'What may we hope for?' And also for those who remember that the root of the word 'probability', the Latin *probare*, can mean either 'to prove, approve' or 'to test, put to the proof', and who therefore admit that confidence can only be proved by putting oneself to the test in a real-life situation.

Since confidence is about more than self-knowledge and knowledge of others, more than admiration, contemplation or familiarity, since it must be based on real reciprocity and the possibility of jointly accepting either success or failure, in a relationship that involves both a hierarchical chain of command and a collegial team spirit, confidence is closely associated with authority. Such confidence can be inspired by many sources and under many conditions; there may also be many undesirable consequences when one of these two qualities is absent, insufficient or ignored. The Challenger disaster provided a tragic and by no means unique example: there can be no authority without the establishment of, and desire for, genuine confidence between those who exercise, apply and share it and those who are subject to it. If there is a lack of resources, including in the means of communication and discussion, if the legitimacy of those in authority is questioned or found lacking, if their scientific or technical credibility is considered to be flawed or defective, authority loses all relevance and the possibility of reciprocal confidence goes out the window. As a result, the cursor on the scale of risk threatens to slide too far in the direction of failure and tragedy, rather than towards success and safety.

While profound and personal belief (in French *intime conviction*) never seems to have been considered a cardinal virtue, it could perhaps claim to be one: at minimum, I would say that personal belief is for confidence what precaution is for prudence. Indeed, though it is certainly often found in the technologically-advanced sectors of aeronautics and astronautics, these are by no means the only areas where it is important. In trying criminal cases, French law includes a formal procedure of asking witnesses to express their 'innermost belief': *'Quelle est votre intime conviction?'* The witness is then expected to convince the jury by his sincerity, knowledge and convincing evidence. The listeners assume that this firm belief or 'conviction' will be profoundly personal in the sense that it will represent the witness's innermost and most deep-seated feelings, perhaps even the most secret. This innermost conviction needs to be more than a mere impression which would be too general, hasty or superficial; on the contrary it requires close and conscious reflection, a rigorous analysis of each aspect of the issue, every item of evidence, every fact submitted by the prosecution and the defence. To call on, to ask for, indeed to insist on knowing someone's innermost feelings is more than a formality, more than a bureaucratic procedure, more than a cognitive attitude; it associates the quest for the good of the individual and society with a kind of modesty when faced with the hard facts, the imperfection of human knowledge, the responsibility involved in judging another, and the serious consequences of one's decision. Because it combines thought and action, calling on one's innermost feelings has a genuinely ethical dimension, with one essential characteristic, even if this may seem paradoxical at first sight, that of collegiality. It should be remembered and admitted that this is a genuine constraint, as it requires time, patience and an effort to find agreement. And to what end? To reach a verdict, in other words 'to say truly how I as a human being judge the situation'.[3]

While the French justice system regularly uses *'intime conviction'* as a practical consideration, it is also a very real 'virtue' of aeronautical and astronautical engineers and technicians as they develop innovative technologies and procedures. After a series of studies and partial tests, after reviewing all the flight parameters for a prototype

---

3. See Jean-Marie Fayol-Noireterre, 'L'intime conviction, fondement de l'acte de juger', *Informations sociales* 127 (7/2005): 46–47.

aircraft or rocket, and after examining the weather conditions, the moment finally comes to launch the new creation into the skies. Those to whom this decision is entrusted must have the '*intime conviction*' that it is possible, appropriate and, at this point in the process, necessary to make this first flight. The demands, constraints and ethical considerations they face can be described using the same terms as in the judicial context: method, reflection, humility, collegiality, etc. And of course the people concerned also need to be as completely as possible in control of the moment when everything changes, the moment of "catastrophe" in the Greek sense and as used by Rabelais, the moment bounded by the experience of the past and the expectation of the future, this 'interval in time which is entirely determined by things which are no longer and are not yet' (Hannah Arendt). This is the moment of risk *par excellence*, of calculated hope and, when all is said and done, of something very like the 'moment of truth'.

Prudence and precaution, confidence and conviction: the exercise of the virtues is binding upon those who explore space. Perhaps I should add the virtue of courage, in the sense proposed by Friedrich Nietzsche:

> Do you have courage, oh my brothers? Not courage before witnesses, but the courage of hermits and eagles that even the gods cannot see? [. . .] He who has heart knows fear, but conquers fear, sees the abyss but with pride. Whoever sees the abyss but with eagle's eyes, whoever grasps the abyss with eagle's talons, he has courage. (*Thus Spoke Zarathustra* IV, 73, Sect. 4.)

We need this courage, just as this caution and this confidence, to advance a step, also modest it may be, above the abyss revealed by space exploration.

12
Casting the Dye

In his *The Civil Wars*, Julius Caesar says nothing of this. However, Plutarch, Suetonius and Appian gave him the words which would be etched in history, in the collective writings of Latinists and, eventually, in the pink pages of the *Larousse* dictionary: '*Alea jacta est*—The die has been cast.' It was at the moment of crossing the Rubicon, the shallow river running through the Po Valley, that the Roman Senate, in the middle of this last century BC, constructed a sort of symbolic wall: anybody climbing over it to threaten Rome at the head of an army, a legion or a simple cohort would be declared guilty of parricide and sacrilege. Caesar knew it. He must have hesitated for a moment before crossing the Rubicon with his troops in an attempt to overthrow Pompey and take power. His act was no less perilous than that of Leonov.

Without asking too many questions about the historic veracity of the use of this lapidary formula by the Proconsul, it would be better, by learning of its Greek origin and the slightly different meaning this gives it ('That the die was cast!'), to ask how it should be interpreted. Does the man that speaks thus abandon himself to chance as this word has its roots in the Arab word designating the same casting of dice? In other words, does he let himself be swept along by the rushing and untameable flow of events that have brought him to the banks of the Rubicon without knowing the outcome of his military attempt, without hoping to be able to retrace his steps or of crossing it to escape? Or did the ambitious Proconsul choose to speed up history, to take life into his own hands, to run the risk of offending the gods and, more mundanely and immediately without doubt, confronting the Senate? Between the religious reasoning of the era and personal

ambition, it was hardly easy to determine to which of these to attri-
bute the exceptional destiny of Caesar. Even an historian retracing
the journey of a man, a community or a country through time, is
faced with choices to make, of reaching the point of no return.

Events such as these jumps, crossings of boundaries and explorations referred to take on a singular tone which very often struggles to
describe the mysterious alchemy of freedom and desire, chance and
necessity. And yet we dare to tackle this task, as it leads us to describe
an essential element of the act of exploration. To do so, I want to refer
to the definition of chance, devised by Antoine Cournot in his work
The Theory of Chance and Probability, published in 1843. Inspired by
Aristotle's *Metaphysics*, he describes chance as 'the combination or
the encounter of other events pertaining to mutually independent
series are those which we name fortuitous events or results of chance'.
This definition is challenging: how can we ensure that two sequences
of events are effectively independent of each other? The operation is
perhaps more complex than initially thought. The basis for the experience of thought as defined by Albert Einstein, and known as EPR,
performed much later by Alain Aspect, questions the existence of
links that may exist between particles yet begin outside our normal
experience of reality. Another equally recognised way of looking at
chance was referred to by Henri Bergson in *Les deux sources de la
morale et de la religion*, published in 1932: 'An enormous tile, torn
off by the wind, lands on and knocks out a passer-by. We say it is
chance. Would we say the same if the tile had simply smashed into
the ground? Chance is therefore the mechanism that acts as if it were
intentional.' It is difficult to deny that the two events are not linked:
the event that leads this individual on a walk and that which leads to
the tile falling off the roof . . . unless we imagine a scenario that puts
this future victim as the owner of a house, whose roof, in need of
repair, requires the work of a roofer, a scenario that would envisage
the impossibility of getting hold of this roofer even though the storm
threatens, which would then imagine the owner's alarm, his decision
to go out to try to find someone to repair his roof . . . before being hit
by one of his own tiles. Putting this rather farfetched hypothesis aside,
let us consider the lesson of these two philosophers. If we stick to the
theme of this essay, that is exploration, of plunging into the unknown,
we can *a priori* see the series of events as unrelated, in the same way
as the places which are known and unknown to the explorer. If such

was not the case, if there was a link, albeit tenuous, between them, a link from which we exclude however that of the imagination, could we still talk of exploration? Would this not be more a journey of discovery, of rediscovery or simple leisure trip?

If we adhere to Cournot's definition of the meeting of two completely independent worlds, it is important to recognise, to expect and even to demand a type of 'improvisation', of chance in the exploration venture. In doing this, it is advisable to disassociate chance from the chaotic and confused dimension that is often wrongly associated with it. Let us consider the game evoked by Caesar on the banks of the Rubicon: one die is a polyhedron but only cubic and each of its six sides is different from the other five. In other words, one die offers a given number of possibilities, among which one and only one can occur. This is a good way to interpret eventuality. Caesar perhaps made a choice in defying Rome. He refused to retreat and chose confrontation; he configured a new area of military and political possibilities, definitely more dangerous, but also doubtlessly more limited and maybe easier to understand. The die was cast and it was Caesar who took hold of it and decided to throw it.

Like him, an explorer leaves nothing to disorder. He knows the odds involved in the game from which reality will soon emerge once the die is cast and his world meets with an unknown world. He knows the moment is coming and remains for as long as possible in control of the arrival of its carriers, its procession and its craft. Is this the right moment or *kairos* as the Greeks said? The explorer hesitates: there is still time to abandon the mission, to remain in the wings on the bank, on familiar and firm ground . . . The explorer is not a tile which, blown by the wind, comes loose from the roof, or an asteroid which, set in its orb, inexorably approaches Earth. For a brief time, he is still in control of the game, the organiser of the encounter. His men and his beasts shuffle along, all fascinated by the emptiness of the unknown, the point of no return. His sails are out, his engines are rumbling, his instruments are set. Yet he waits still. Never has he felt so powerful, so big, so in control of his own existence. Never has he felt so close the intoxicating and reeking breath of madness. Will he give the signal? Will he press the button? Will he open the gateway? Will he open the eye of his telescope? Will he launch the powerful particle beam? Is it too soon or is it already too late? Never will man feel so alone, in space and time, than when confronting the unknown. But also so free.

13
The Earth does not Move

Let us be bold enough to put forward the following hypothesis: the most unexpected, the most paradoxical, but maybe also the most assured of consequences of half a century of exploration and conquest, exploitation and use of space, would be to have diminished, weakened and even rendered vain the spirit of exploration. This hypothesis is without a shadow of a doubt openly daring and iconoclast, and there seem to be plenty of reasons and evidence to disprove it. It is also necessary if we are to limit ourselves to the unique world of astronautics, with so many study programmes looking at the astronomy of the universe and sending interplanetary probes. It might be to look for exoplanets or to scan the depths of space and time to better discern and understand the first cries of our universe. It might be to observe the planets closest to Earth to study their geology and aptitudes to enable and even to provoke the emergence of biological forms. Or, it might be to gather fundamental building blocks, reaching asteroids to study them and adding them to our galaxy archives. There are numbers of reasons why surely we must continue to explore this immense unknown world that surrounds us, to which we belong and from which we ourselves have emerged. Those who develop future space programmes, those who are responsible for informing others never cease to awaken our curiosity, to popularise exploration, which they very often portray as obvious to better justify it. Even so, I am not prepared to abandon this unsettling and frustrating hypothesis so quickly and easily. Despite all the efforts of researchers, their political support and their communicating, regardless of technological challenges and financial difficulties encountered by these projects and these exploration programmes, is it not honest

and realistic to ask whether our era is missing the taste of curiosity
and the enthusiasm for exploration, but also the interest for conquest,
that reigned within our cultures and our societies in former times? I
would therefore like to study the hypothesis according to which one
of the reasons for this lack of inspiration can be paradoxically found
in the extraordinary and perhaps unexpected space successes of the
1960s and 1970s, which took us on a path back to Earth!

These two decades saw mankind reach the Moon and gave twelve
Americans the chance and courage to accomplish the incredible
odyssey. We fulfilled the dream prophesied by Tsiolkovski and briefly
left our cradle. But what were we in a hurry to do once in space, even
before the Apollo missions and from the date of Yuri Gagarin's inau-
gural flight in April 1961? As I have already said, we made a U-turn
and turned our thoughts to Earth . . . One man, long before the work
of modern astronautical pioneers, imagined it would be this way:
Lucian of Samosate, in his *Icaromenippus*. Fifteen centuries before the
imaginary journey of Cyrano de Bergerac, eightteen centuries before
the real of Neil Armstrong and Buzz Aldrin, he succeed to join the
Moon. When he arrived, he sat down and, forgetting the astronomical reason for his risky journey . . . he looked down upon Earth! This
premonition penned by the second Century satirist was strange: the
first men to orbit the Moon, followed by those who walked on it, did
look down upon Earth. What souvenirs did they bring back from
their missions? They brought back rocks from the Moon and images
of the desolate plains on which they landed but, above all, they told
us of their unforgettable feelings about the Earth and incomparable
views of our planet 'blue like an orange' (Paul Eluard) in its black cas-
ing dotted with stars. It is at this point that the two stories—that of
Menippus and our own story—diverge. Having grown weary of seeing humans like ants busying themselves carrying a grain of wheat, a
bean husk or a blade of straw; the adventurous ancient traveller took
flight once again, determined this time to head for paradise. However, the astronauts returned to Earth ready to defend future space
exploration programmes but also concerned for the future of our
planet. The first steps of space explorers quickly put us on a path back
to Earth.

This new view of Earth came just at the right time. Rachel Carson
published her book *Silent Spring* in 1962 and, ten years later, the first
international United Nations conference on the environment took

place in Stockholm. However, should ecological concerns be considered as the only explanation for this dwindling interest in the space exploration venture? Without denying the influence of this coincidence between the space and environmental movements, I believe that the first decade of space adventure at least fulfilled the prophecy of Tsiolkovski confirmed by Edmund Husserl's vision.

In 1934, the German philosopher wrote a study that was published much later and the title of which must be given in full: 'Overthrow of the Copernican theory in the usual interpretation of a world view. The originary ark, earth, does not move. Foundational investigations of the phenomenological origin of corporeality of the spatiality pertaining to Nature in the first sense of the natural sciences.' In no way Husserl questions the scientific input of Nicolaus Copernicus and the astronomer born in the sixteenth Century; he simply questions his position with regard to Earth from a phenomenological point of view. The Earth does not move, the philosopher articulated, because it is not, for humans, a vessel like a ship, an aircraft, a land or even a star can be. For us as humans, the Earth is not a body, but a ground, phenomenologically immobile even though it moves in the cosmic vacuum. It is a ground for the simple and good reason that all other realities we know or can know relate to it, depend on it and find their origins in it. Husserl speaks in particular of Earth as the original ark.

What if, I reformulate my hypothesis, the first decade of space adventure supported Husserl's theory? What if, above all, it made us take stock of our terrestrial condition, giving us a global vision of our own planet within the universe? The Earth is not an anonymous celestial body among thousands, millions and billions of others; it is not only a cradle that we can or must leave one day. It is our ground, 'Mother Earth', for many years to come, to which we return from our interplanetary journeys and from which we look toward the cosmic vacuum. That is what five decades of space conquest have taught us. It is not astonishing therefore that we have (momentarily) lost the necessary impetus, the visceral desire to look elsewhere, to explore a world that has become sky-high, empty and without horizon. The Earth has, for a time, become, whether you like it or not, the main destination of the human odyssey. Even so, we have not lost our unfailing and innate curiosity.

14
The Greater Earth

By affirming that 'The originary ark, earth, does not move', Edmund Husserl makes a distinction between the body or multiple bodies and the single ground. Remember also that the Earth is now considered a ground. For the time being and maybe for a long time to come, we should not dream of a new Earth as our predecessors were able to imagine, seek and, sometimes, conquer new worlds, eventually abandoning the old ones. The step is too big, the jump too dangerous, to reach a celestial body that could be inhabited by humanity. It would seem more logical to push back Earth's space frontiers or to define new frontiers, to cross them gradually, to discover and extend this domain which, ten years ago, was baptised 'the Greater Earth'.

To spark new interest in space exploration, it is perhaps not necessary, in the space of reasonable possibilities that is today our own, to accumulate light years, or to resort to forecasts which end up resembling works of science fiction. The next destination or the next goal, I should say, is located one million kilometres above our heads. This new frontier borders *Greater Earth*. In this space territory or area that surrounds the Earth, most future space activities could theoretically take place, without requiring overly high energy costs. A space platform is either naturally held there by the Earth's gravity, or maintained in this terrestrial neighbourhood at the cost of modest propulsion manoeuvres. The L3 Lagrangian point in the Sun-Earth system is one example. Platforms and bases developing in this domain, instead of being trapped at the bottom of gravitational potential wells, can be easily accessed: from the Earth, the propulsive evaluation is, within one or two km/s, the same as that of placing into geostationary orbit. The propagation time of radio electric waves does not exceed three to four seconds, signifying that exchanges with the Earth may be naturally continual and cooperative, between humans and/or instrumental systems, without imposing autonomy constraints, inherent to Mars missions for example. Finally, in the event of human presence on board, the time required to return to Earth is a few days, comparable to that of lunar missions. The next logical step for space venture, for the reasonable use of space by our civilisation could therefore be found in the exploration, conquest and development of this Greater Earth.

Is the horizon of this Greater Earth linked to utopia, inspiring dreams but without a real future? A long time has passed since Gerard O'Neill imagined space colonies known not for their fictional dimensions or their populations but rather for their excessively idyllic character. O'Neill did not only imagine planting vegetable gardens in these famous spheres in order to provide sustenance and a source of moral fibre for humans but also described how it would be possible for birds and butterflies to live there but banning entry to rats and mosquitoes![1] Greater Earth will probably not be utopic, because it is already being created in the area closest to our planet, the front line being formed by satellites along the geostationary belt. To such an extent perhaps that it would be better to abandon the term 'outerspace' which designates the area of orbits around the Earth, even if this usage is commonly employed in international treaties and agreements. This area near to Earth seems to be seen less and less like a slightly neglected suburb of our planet and increasingly as a component element that is essential to Earth's future. As the lord who imposed his law in the past, his 'ban' on a territory of a league or several leagues, humans scarcely hesitated or delayed in defining this area of space surrounding the Earth as a 'suburb', in imposing on it their own jurisdiction and in declaring it a common asset of mankind: everyone can, depending on his capacities, go there, put machines there, practice an activity there, both for- and non-profit. Mankind must simply accept common governance and cannot claim any ownership. When we talk about the importance of circumterrestrial activities with regard to the current and future human activities of our planet, is it not time to start talking about Greater Earth as this

---

1.   Cf Gerard O'Neill, *The High Frontier. Human Colonies in Space* (New York: William Morrow & Company, 1977).

territory encircled by geostationary satellites and to respect it in the same way as we would good old Mother Earth?

Accomplishing the step of the Greater Earth would reinforce the awareness touched upon and analysed by Husserl: humanity should be more aware of its singular origin on Earth and, consequently, of its unity as a species. The 'we', experienced during the Apollo missions, would be experienced even further. Space exploration would therefore be social, political and human: our species would undertake the exploration of the noosphere.

In December 1943, when Vladimir Ivanovich Vernadsky defined the concept of "noosphere", *Big Brother* and satellites only existed in the imaginations of writers and in the minds of engineers; only the German V-1 and V-2 missiles allowed people to imagine that astronautical techniques in the future could offer humanity new horizons if humanity survived the violence weakening societies across the entire globe. It was at this time that the Soviet scholar wrote an article which was published one year later in *The American Scientist* under the title: 'The biosphere and the noosphere'. He stated: 'The word "noosphere" derives from the Greek *noos*, and means mind, while *sphere* means globe, an envelope around the Earth.' And he continued: 'Humanity, as living matter, is inseparably linked to the material and energy processes of a geological envelope specific to Earth—its biosphere. Humanity shall not be physically separated from the biosphere for a single minute.' Vernadsky saw the noosphere as the final stage in the development of the biosphere due to strictly geochemical processes. He wrote:

> We are entering the 'noosphere.' This new geological process took place in troubled times with a world at war. However, the most important thing is that our democratic ideals are applied to basic geological processes, to laws of nature and to the noosphere. Thus, we can face the future with confidence. It is in our grasp. We must not let it slip away from us.

Imagined by Vernadsky more than seventy years ago, the noosphere is today witnessing technological development and anthropological exploration. A vast still-unknown territory that surpasses or rather overwhelms all previous human experiences of interpersonal relations. As Jacques Julliard predicted on 21 November, 1992, following a silent protest against president Milosovic's ethnic cleansing policy, 'tomorrow, they cannot plead ignorance, they cannot say that it wasn't possible'. Never was this 'tomorrow' so close to our today; never has humanity as a whole been faced with such an unknown, without having even left Earth, the original ark, and having found an essential horizon.[2]

2. Jacques Julliard, 'Nous ne pourrons pas dire que nous ne savions pas', *Esprit* (Janvier 1993): 138–139.

15
Plea for a New Horizon

As this essay on space exploration draws to an end, it is natural for the reader to raise questions and ask me what I was hoping to achieve. I have painted a picture of space adventure which, following the resounding success of the first fifteen years, then, in my opinion, entered a period of banality and failure. As a consolation, I advocate a Greater Earth which seems to involve the colonisation of an area of space already explored and known. Do I therefore agree with those who consider space venture, from an exploration aspect (as the usefulness of its 'satellite' aspect, with regard to Earth, is not denied) to have been simply a chapter in humanity's history, a remarkable chapter certainly but one that should come to an end? In short, is this my personal belief?! I hear this comment, this critic and I would like to answer your questions as fully as possible.

In a previous chapter, I wrote that, as I see it, humans explore just as they breathe. To return to this deliberately exaggerated observation, I would also say that everything that constitutes a difficulty, a restriction in man's penchant for exploration, in the innate desire and in its realisation, can play a part in stifling it. I am not referring to physical difficulties. I am talking about cultural and spiritual aspects that characterise and differentiate human beings and, more widely, the human species. This premise must clearly be handled with care, in other words using the critical and rational mind that belongs to our species and its cultures. We must learn from the past, from the major explorations of modern times, since the sixteenth Century. We must also learn from the shorter period of time when the ideology of progress prevailed unchallenged.

The notion of noosphere that I marked as useful in the construction of Greater Earth is incontestably linked with very notion of progress: the noosphere can be considered as the aim of the processes which, since the beginning of our universe, appear by ensuring its transformation and development. Due to its age and firm foundations, should we accept all the forms, all the costs and all the consequences? Due to the terrifying battle fields, concentration camps, mass tyranny, worrying levels of waste and unacceptable pollution of the twentieth Century, the very idea of progress, and, above all, its dictatorship was called into question. However, I do not believe that the desire for progress in the way we are and in our actions, good and bad, can be eradicated from human nature. I say it again: to decide that exploration, one of the most human forms of progress, is now excluded, prohibited from the human cultural sphere would be catastrophic for our species and would signal its end. I think that now would be a good time to add an element that I have not yet introduced in my reflexion.

Without it being possible to prevent faux-pas, mistakes and errors that indubitably go hand in hand with human venture, no matter how well founded, thought out or rational they are, it is important to care about the introduction and maintenance of an essential dimension of the human odyssey: that of the sacred. This word can be overwhelming due to its usage through the history of mankind, still trailing shadows, ghosts and darkness in its wake. It is important in this case to ignore the most obscure and bloody aspects of sacrifice and to think only about the most enlightening aspects of the sacred.

What is set aside and separated, invested from an intangible value and an inviolable character is declared sacred. The *fanum* area of a temple is separated from the *profanum* area which surrounds it. Holy of Holies with many names, prohibited from the common worshipers, separated by a threshold, itself sacred, that only priests can cross. Still declared sacred are those who feel the divine, to serve it or through a particular act. In many religions, new mothers find themselves momentarily sacred and blessed following their experience of the mystery of life itself and of birth. She cannot have normal contact, an ordinary exchange with beings, even those who are the most familiar; for fear that they will be soiled. She must be purified, like a vase that has been used for a religious rite. Also, one who has committed a crime against the gods or the State is also declared sacred. He is shut away, separated from the religious or social group. Only his death can restore order and peace to the world. It is not surprising that the galactic centre was, perhaps the first, to be declared sacred. After all, is it not both frightening and captivating for humans, does it not have the *tremendous* and *fascinating* traits as described by the German philosopher Rudolf Otto? This sacred character did not immediately disappear with the beginnings of space exploration. Astronauts and cosmonauts themselves were the subjects of a process of sacralisation, albeit unwittingly. From the colour of their clothes (the white of purity, the red of sacrifice) to the ascetic rigour of their training, the strict selection process and the aura of their fame – everything was in place to make them into heroes and sacred people. Being made sacred again when they returned from space, the American astronauts plunged into the waters of the Pacific Ocean, as if in a bath of purification, and faced their final trial of quarantine. The sky, the cosmos, which was now space, had not lost any of its power; it had even made a small portion of humanity sacred!

Humans need the sacred, not only to adore it from far and in times of worry, but also to be attracted by it, compelled to do the best they can in order to attain it. The very nature of sacred allows it to be transgressed! This idea may appear surprising and is indeed surprising: the process of making a reality sacred involves the need to specify conditions for its transgression, whether it is an era, a rite, etc. At the centre of the temple of Jerusalem, the Holy of Holies was the Jewish faith's sacred place *par excellence*; a place no one could enter, except the High Priest, once a year, to pronounce the name of God which otherwise could not be uttered. And so it is for all Holiest of Holies, for all places established or recognised as sacred by humanity: they must be able to be transgressed to have a true influence over those who respect it. Was one of the most deeply moving sentences uttered by a Palestinian Rabbi two thousand years ago not: 'The Sabbath was made for man and not man for the Sabbath'?

It is because of this that I believe it is essential to defend and encourage exploration within our human societies: exploration is made for man. It demands and allows us to determine the meaning of sacred that is not a taboo or a protection but instead a driving force, a projection. It requires and allows us to dream of new horizons and to

expand our internal and external desires. Does exploration not honour this? Does it not ensure the dual act of shrinking the universe
to fit human dimensions and of expanding human dimensions to fit
those of the universe? Thus, exploration is an essential element of the
humanism of space that I champion with all my heart[1].

---

1. Cf Jacques Arnould, *Icarus' Second Chance. The Basis and Perspectives of Space Ethics.* (Wien/New York: Springer, 2011), 190.

References

Arnould, Jacques. *La Terre d'un clic. Du bon usage des satellites* (Paris: Odile Jacob, 2010).

Arnould, Jacques. *Icarus' Second Chance. The Basis and Perspectives of Space Ethics* (Wien/New York: Springer, 2011).

Brunier, Serge. *Impasse de l'espace. À quoi servent les astronautes?* (Paris: Seuil, 2006).

Cosgrove, Denis. *Apollo's Eye. A Cartographic Genealogy of the Earth in the Western Imagination* (Baltimore & London: Johns Hopkins University Press, 2001).

Dator, James A. *Social Foundations of Human Space Exploration.* (New York: Springer-ISU, 2012).

Koestler, Arthur. *The Sleepwalkers: A History of Man's Changing Vision of the Universe* (London: Pelican, 1968).

Lafferranderie, Gabriel. 'Espace juridique et juridiction de l'espace', in Esterle, Alain (ed.). *L'Homme dans l'espace* (Paris: PUF, 1993).

McCurdy, Howard E. *Space and the American Imagination* (Washington & London: Smithsonian Institution Press, 1997).

O'Neill, Gerard. *The High Frontier. Human Colonies in Space* (New York: William Morrow & Company, 1977).

Pons, Walter. *Steht uns der Himmel offen? Entropie-Ektropie-Ethik. Ein Beitrag zur Philosophie des Weltraumzeitalters.* (Wiesbaden: Krausskopf Verlag, 1960).

Wolfe, Tom. *The Right Stuff* (New York: Farrar-Straus-Giroux, 1983).

Index of Names

A

Aldrin, Edwin Eugen « Buzz », 9, 10, 14, 20, 37, 49, 68.
Alighieri, Dante, 1.
Amundsen, Roald, 21.
Apollinaire, Guillaume, 4.
Appian, 63.
Arendt, Hannah, 5.
Armstrong, Neil, 9, 10, 13, 14, 20, 27, 40, 68.
Auger, Pierre, 6.

B

Beliaïev, Pavel, 19.
Bergerac, Cyrano de, 2, 3 68.
Bergson, Henri, 64.
Bernard, Claude, 45.
Blanchot, Maurice, 18, 19, 20.
Bonestell, Chesley, 4, 11.
Bowie, David, 13.
Braun, Wernher von, ix, 4, 37.
Breuil, Henri, 26.
Brunier, Serge, 15, 41, 79.
Burgat, Florence, 49.

C

Caesar, Julius, 63, 64, 65.
Cameron, James, 51, 52.
Carson, Rachel, 69.

Chrétien, Jean-Loup, 13.
Clarke, Arthur C, 2, 3, 12, 51.
Columbus, Christopher, 26.
Copernicus, Nicolaus, 26, 69.
Cournot, Antoine, 64, 65.
Cuarón, Alfonso, 20.

D

Darwin, Charles, 25, 26, 27.
Disney, Walt, 4, 10.

E

Einstein, Albert, 64.
Eluard, Paul, 68.
Ewald, François, 31.

F

Flammarion, Camille, 2.
Fontenelle, Bernard de, 3.

G

Gagarin, Youri, 5, 37, 45, 50, 68.
Galilei, Galileo, xii, 4.
Gazenko, Oleg, 46, 47.
Goddard, Robert, 4.
Godwin, Francis, 1.

H

Hadfield, Chris, 13.
Hegel, Georg Wilhelm Friedrich, 17, 19.
Hobbes, Thomas, 21, 22.
Husserl, Edmund, 69, 71, 73.

J

Jacob, François, 6.
Julliard, Jacques, 73, 74.

K

Kant, Immanuel, 29, 54, 59.
Kennedy, John F, 4, 5, 35.
Kepler, Johannes, xii, xiii.
Korolev, Sergei, 37.
Kranz, Eugene F, 32.
Kubrick, Stanley, 12.

L

Lafferranderie, Gabriel, 42.
Lansdorp, Bas, 35.
Latour, Bruno, 57.
Leonov, Alekseï, 18.
Ley, William, 4.
Loti, Pierre, xiii.
Lowell, Percival, 2.
Lucian of Samosata, 1, 3, 68.

M

Magellan, Ferdinand, 26.
McCurdy, Howard, 3.
Mermoz, Jean, 26.

N

Nansen, Fridtjof, 21.
Nietzsche, Friedrich, 61.
Nixon, Richard, 9.

O

O'Neill, Gerard, 72.
Otto, Rudolf, 77.
Ovid, 3.

P

Pascal, Blaise, xiii.
Plutarch, 63.
Poe, Edgar, 2.
Pons, Walter, 17, 18, 19, 20, 52.

R

Rémond-Gouilloud, Martine, 55.
Rozier, Jean-François Pilâtre de, 50.

S

Saint-Exupéry, Antoine de, 26.
Schiaparelli, Giovanni, 2.
Shepard, Alan, ix, 45.
Simmons, Dan, 11.
Socrates, 17, 19.
Spielberg, Steven, 52.

T

Tsiolkovski, Konstantin, 4, 5, 41, 68, 69.

V

Vernadsky, Vladimir Ivanovich, 73.
Verne, Jules, v, 3, 11.

W

Wells, Herbert G, 2.
Wolfe, Tom, 48, 49, 53, 79.

Y

Yeager, Charles Elwood 'Chuck', 48, 49.

CPSIA information can be obtained
at www.ICGtesting.com
Printed in the USA
FFHW021526031019
55349413-61092FF